Matrix-Based Product Design and Change Management

Dunbing Tang · Leilei Yin
Inayat Ullah

Matrix-Based Product Design and Change Management

Springer

Dunbing Tang
College of Mechanical and Electrical Engineering
Nanjing University of Aeronautics and Astronautics
Nanjing, Jiangsu
China

Inayat Ullah
College of Mechanical and Electrical Engineering
Nanjing University of Aeronautics and Astronautics
Nanjing, Jiangsu
China

Leilei Yin
College of Mechanical and Electrical Engineering
Nanjing University of Aeronautics and Astronautics
Nanjing, Jiangsu
China

ISBN 978-981-10-5076-3 ISBN 978-981-10-5077-0 (eBook)
DOI 10.1007/978-981-10-5077-0

Library of Congress Control Number: 2017943171

Printed on acid-free paper

This Springer imprint is published by Springer Nature
The registered company is Springer Nature Singapore Pte Ltd.
The registered company address is: 152 Beach Road, #21-01/04 Gateway East, Singapore 189721, Singapore

Contents

Chapter 1
Product Design as Integration of Axiomatic Design and Design Structure Matrix

1.1 Introduction and Synopsis

Axiomatic design (AD) is a prescriptive engineering design theory that provides a systematic and scientific basis for making design decisions. In AD, two axioms give design teams a solid basis for formalizing design problems, conceptualizing solution alternatives, eliminating bad design ideas during the conceptual stages, choosing the best design among those proposed, and improving existing designs [1, 2]. One of the findings to date is that one of the limitations of AD is its concentration on the architectural design, at the expense of the system design context. As a result, certain factors and constraints, such as cost, time, and physical integration are not catered directly by the axiomatic model [3, 4]. As a structured modeling method, more recently the design structure matrix (DSM) model has been regarded as a good roadmap of system design [5]. DSM is a good tool mapping information flow and its impact in product development processes; and it can represent visually the network of interactions among development activities or design objectives and facilitates analysis of these interactions [6–11]. The process of DSM construction involves understanding, tracing and capturing the interaction relations among system elements, which actually is a process of system level knowledge creation and recording. Knowing system interactions in the product development process is critical for project management and decisions. System interactions mainly refer to the inter-relationships existing between element types during the product development cycle. Capturing inter-relationships between affiliated system elements is necessary if the product engineering is to be handled efficiently and effectively. Currently, DSM has been a common means to represent the system interactions. Although DSM is powerful on the analysis of design interactions, it is less effective in innovative design. One reason is that it is not possible to obtain a DSM for a new product that has never been designed before. Another reason is that DSM fails to record explicitly the reasons for the system interactions.

D. Tang et al., *Matrix-Based Product Design and Change Management*,
DOI 10.1007/978-981-10-5077-0_1

There has been evidence of attempts to link the design matrix (DM) of AD and the DSM. Dong and Whitney present a technique to obtain a DSM from a square DM [3], and Guenov and Barker improve such technique to derive an 'architectural' DSM from a DM, which may not be square [4]. This research considers that AD and DSM can be integrated in a complimentary manner and captures and models the underlying logic of integration of both. In Sect. 1.2, a brief review and analysis of AD and DSM is presented. In Sect. 1.3, the logic of integration between AD and DSM is interpreted; and cursive interaction mechanism between AD's DM and DSM is detailed, which is the backbone of integration of AD and DSM. In Sect. 1.4, the related system development with a case study is given to illustrate the proposed integration method; and Sect. 1.5 concludes the chapter.

1.2 Axiomatic Design Versus Design Structure Matrix

1.2.1 Advantages and Disadvantages of Axiomatic Design

The underlying hypothesis of the AD is that there exist fundamental principles that govern good design practice [1, 2]. The main distinguishable components of the AD theory are domains, hierarchies, and design axioms. The AD method proceeds from a high level of abstraction to detailed design elements (i.e., from meta-product systems to assemblies to components to features to parameters). This activity of definition and detailing produces a prescriptive model of design hierarchy for the design entity in each of three domains: functional, physical, and process. The decisions that are made at higher levels affect the statement of the design definition at lower levels. That is, the design team goes through a process whereby they zigzag between domains to decompose the design problem. At a given level of design hierarchy, there exists a set of function requirements (FRs) defined as the minimum set of requirements needed at that level. The definition of FRs depends on the solutions with design parameters (DPs), namely, defining acceptable FRs may involve several iterations. Before zigzagging FRs, the corresponding hierarchical level DPs shall be selected. Once a corresponding DP can satisfy a FR, a focused FR can be decomposed into a set of lower-level requirements. The design team will develop different solutions for each DP to satisfy a FR and select the best alternative at each hierarchical level according to axiom 1 "Maintain the independence of the functional requirements" and axiom 2 "Minimize the information content of the design". The axiom 1 states that the DPs and FRs are related such that a specific DP can be adjusted to satisfy its corresponding FR without affecting other FRs. The axiom 2 states that the independent (uncoupled) design alternatives that minimize the information content are the best.

Once a set of FRs has been formulated and feasible sets of DPs have been synthesized at a given hierarchical level, the two design axioms are applied to evaluate the proposed design concepts. Guided by the axioms, the design team should be able to conceive, select, and optimize the "best," even at the conceptual stages. Application of the functional independence axiom (axiom 1) can be described in terms of the DM. A DM prescribes the relationships between the FR array and the corresponding DP array at the same hierarchical level, which can be described as:

$$\{FR\} = [DM]\{DP\} \tag{1.1}$$

$$[DM] = \begin{bmatrix} A_{11} & A_{12} & \mathrm{L} & A_{1n} \\ A_{21} & A_{22} & \mathrm{L} & A_{2n} \\ \mathrm{M} & \mathrm{M} & 0 & \mathrm{M} \\ A_{m1} & A_{m2} & \mathrm{L} & A_{mn} \end{bmatrix}$$

$$A_{ij} = \partial FR_i / \partial DP_j \quad (i = 1, \ldots, m \quad \text{and} \quad j = 1, \ldots, n) \tag{1.2}$$

Depending on the type of resulting DM, three types of designs exist: uncoupled, decoupled and coupled. Regarding the uncoupled design, each FR is satisfied by exactly one DP, and the resulting DM is diagonal. Axiom 1 is fully satisfied with the uncoupled design. The decoupled design occurs when the DM is lower triangular, which means that a sequence exists, whereby adjusting DPs in a certain order, the FRs can be satisfied. The DM of a coupled design contains mostly non-zero elements and thus the FRs cannot be satisfied independently. Although AD has been proved to be an excellent design method by real applications [12–15], some shortages are as follows:

- Suh considers that in the case of a product that is new and innovative, the FRs should be defined in a solution-neutral environment without considering any physical solution in mind. This, however, can rarely happen in practice, particularly in complex product environments, where economic considerations dictate maximum possible utilization of mature designs and existing knowledge [4].
- AD guides the designer finding suitable DPs to meet the needs of FRs. But how can the designer know the interactions amongst the DPs, including geometry, spatial layout, interfaces (e.g. logical and physical connectivity), which will decide the quality of design? Meanwhile, for a FR, there may be more than one corresponding DPs (see Fig. 1.1), and several candidate solutions may all satisfy the functional independency axiom. The final solution has to be decided based on the interactions among DPs.

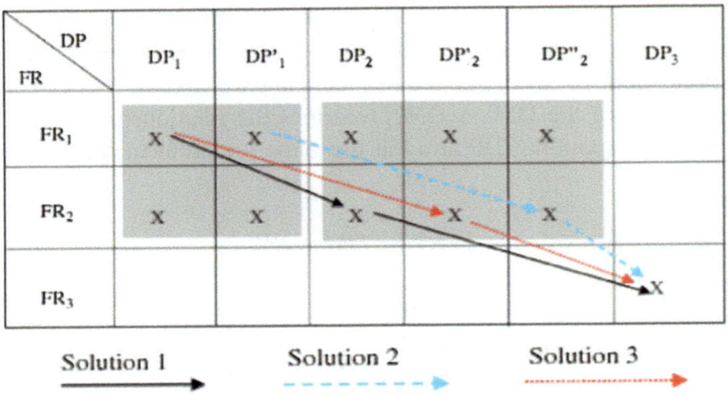

Fig. 1.1 DP candidates for one FR

1.2.2 Advantages and Disadvantages of Design Structure Matrix

The idea of using DSM to represent the system interaction is not new. The DSM method was proposed for system modeling and analysis more than 30 years ago. Steward firstly introduced DSM and developed some algorithms for manipulating the matrix as tools for systems design and analysis [5]. However, it is not until recently that DSM methods started to attract attention for managing the complexity of large engineering systems and complex product development processes [6–11]. A DSM is a matrix representation of a system or a project. System components are listed in the first row and the first column of the matrix. Off-diagonal cells indicate the interactions (i.e. dependency, information flow) between system elements. To construct a DSM, one assigns the individual system element of a domain to the rows and columns of a square matrix. Then, going down the list, if element *b* is an input to element *c*, one puts a mark in the column of *b* and the row of *c* (see Fig. 1.2). One continues until all (known) hierarchical and interdependent relationships have been accounted for. Marks below the diagonal represent forward information transfer which affects the later element. Marks above the diagonal represent information feedback or iteration. The DSM method facilitates minimizing iterations in the process. Partitioning and tearing are always two operations used for this purpose. Partitioning is the process of reordering the DSM rows and columns, so that the DSM is transformed into lower triangular form. Tearing is the process of choosing the set of feedback marks that if removed from the matrix will render the matrix lower triangular. The marks that are removed from the matrix are called 'tears'. Identifying the tears that result in a lower triangular matrix means that the set of assumptions need to be made in order to reduce design process iterations.

In a DSM, the off-diagonal marks are not simply placeholders. Corresponding to each is a specific real issue, involving at least two system elements. The matrix understates the depth of the designers' knowledge for them; each of the marks

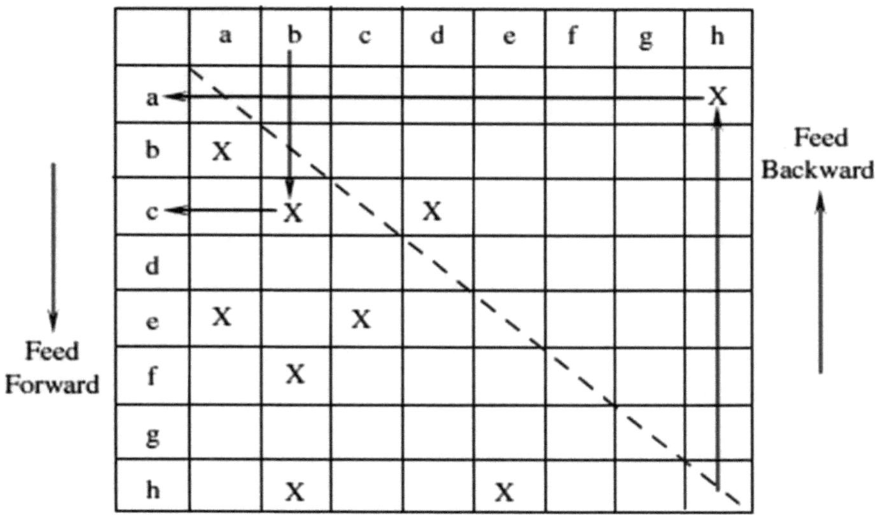

Fig. 1.2 The general DSM form

stands for an existing specific problem or question. DSM has the advantages at recording the interaction relationship between existing product elements. However, for a new product that has never been designed before, it is difficult to make the DSM. At the conceptual design stage, it is also difficult to make the DSM.

1.2.3 The Benefit of Combining Axiomatic Design and Design Structure Matrix

AD is more for creative design and DSM is more for modeling relationships of existing elements of designs. AD is incapable of analyzing the system interactions, which is the greater strength of DSM. DSM cannot cope with creative new design; while AD could cover this shortage. Therefore, the two combined together will be more powerful. It has already been found that AD and DSM could be linked with the conversion of AD's DM to DSM [3, 4]. The reasoning behind these attempts to combine the DM with the DSM was to obtain the design information flow at an early stage of the design, and thus allowing the use of the DSM at the time when the most important decisions about the system and design are made [3]. In combining the two matrices, therefore, the concentration has tended to be on deriving corresponding DSM from the DM, which could lead to the integration of AD and DSM. The logic of integration between AD and DSM, however, is still an open question. In this research, the integration of AD and DSM includes following issues: (1) using existing DSM-based knowledge to accelerate AD; (2) conversion of DM to DSM to get the interactions among DPs at the conceptual design stage; (3) using derived DSM to evaluate the design result of AD from the DPs interaction view; and (4) using derived DSM to conduct project planning at an early design stage. Meanwhile, the intrinsic logic of integration between axiomatic and DSM is uncovered by formal representations.

1.3 The Theory of Axiomatic Design and Design Structure Matrix Integration

The complementary manner between AD and DSM has been found by researchers, whist the conducted research is by no means concluding and profound. The entire exploration of integration between AD and DSM is still in pre-theory stage. One of the major reasons is the lack of a good combination of precise mathematical representation languages and laws governing AD and DSM. The aim of this section is two folds: (1) to establish a basic mathematical formal representation scheme to define the predicates involved in AD and DSM; and (2) to investigate the logic of integration between AD and DSM with the mathematical representation of design objects.

1.3.1 Formulation of Related Predicates

From the AD point of view, product design is started from the customer requirements (CRs), considering various kinds of design constraints (DC) and arriving at a final design solution after an iterative mapping process. The CRs are usually too vague to be tackled directly, and must be interpreted before any attempt on solution. Therefore, a designer associates the CRs with FRs which are engineering language, and tackles the CRs specified in this way. Relation defined in Eq. (1.3) expresses an assertion of such a reasoning step.

$$\exists FR \subseteq \Omega : \lambda_{\text{specify}}(FR, CR) \tag{1.3}$$

where λ_{specify} is the predicate representing that FR can fully specify CR or not and Ω is the function definition domain.

How to find solutions to realize the FRs is the main aim of conceptual design. AD has given axiom 1 to judge the DM and to evaluate the solution candidates. The process of DM construction is actually knowledge-based. In spite of the fact that such knowledge may be very difficult to express, there is still a significant chance for improving our understanding of the DM construction process. Although it may not be possible to capture and represent the inarticulate knowledge in explicit terms, it is interesting to look at how this knowledge is activated and applied. During the mapping process from FR to DP, the designer in the first place considers that the FR

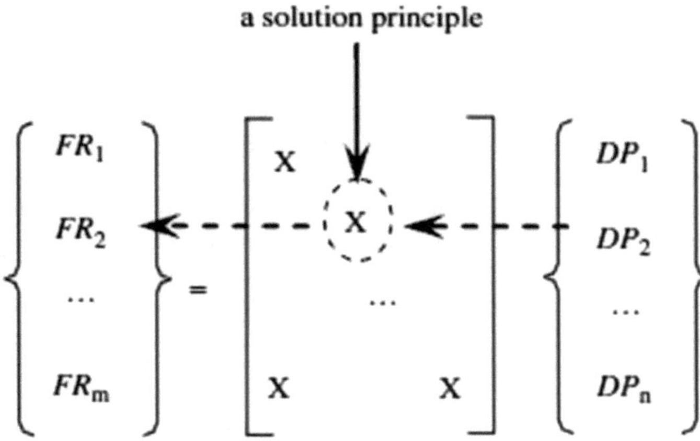

Fig. 1.3 DM construction with solution principles

can be realized though what solution principle (SP), and then conceives the corresponding structure candidates guided by the specified SP. The SP could be physical principle (such Newtonian law of action and reaction), chemical principle, magnetic principle, etc. In other words, the main diagonal elements of AD's DM are associated with various solution principles (see Fig. 1.3).

A specified SP is selected from a domain theory Y, which is a generic, problem-independent knowledge, possibly applicable to different problems. For example, physics is a generic domain theory applicable to design of an elevator as well as a spacecraft. A SP is 'instantiated' for a particular conceptual design, thus creating a usable theory for solving a particular problem. With the specified SP, the mapping from FR to DP can be formulated in Eqs. (1.4–1.6).

$$\exists SP \subseteq \Theta : \lambda_{\text{satisfy}}(SP, FR) \tag{1.4}$$

$$(SP{:}FR \rightarrow DP) \Leftrightarrow (SP \sim DM) \tag{1.5}$$

$$DM_{ij} = \frac{FR_i}{DP_j} | \forall \ i \in [1, \ldots, m], \ \ \forall j \in [1, \ldots, n] \tag{1.6}$$

In Eq. (1.4), we propose to define a problem-solving model as a minimal sub-set of SP that satisfies the explicit function requirements. The relation λ_{specify} would be binary, because it associates a problem-solving model with the current explicit function requirement. Equation (1.5) denotes that the mapping from FR to DP is supported by SP, and the construction of DM is guided by SP.

Based on the constructed DM, AD can evaluate the solution quality according to axiom 1 from the function point of view, but cannot guarantee the design technical feasibility from the physical synthesis point of view. Clearly, a system or product is made of a number of modules. Each module is pieced from a set of DPs, which is a key design decision to be taken for physical design synthesis. A DP set forming a design module may has elements that belong to different design mappings in several hierarchical levels. Feasibility thinking should guard this synthesis activity [16]. Design feasibility here has two aspects: the first is that the set of DPs constituting the module should be able to deliver the FRs hosted in the same module. Second, the relations or interactions among DPs shall be feasible for modular design and the module itself should be feasible as to good properties (such as reliability, safety, manufacturability, assembly, cost, etc.), which means that the constituent DPs module can be synthesized in a manner that meets the required properties.

To check the design technical feasibility, the relations among DPs would be the main focus to be considered and DSM subsequently is capable to deal with the description of interactions among different DPs (see Eq. 1.7)

$$DSM_{ij} = \frac{DP_i}{DP_j} | \forall \ i,j \in [1,\ldots,n] \tag{1.7}$$

It has been noted that for a new product design or conceptual design stage, it is difficult to make a DSM directly as the interaction relation among DPs is vague. Dong and Whitney [3] and Guenov and Barker [4] have addressed the conversion of DM to DSM, which can be defined by

$$T{:}DM \rightarrow DSM \quad \text{or} \quad DSM = T(DM) \tag{1.8}$$

The conversion operator T includes following operations: (i) in each row of the DM chose a dominant entry (X0 in DM) as output variable; (ii) construct a composite matrix to describe the equation relationship between DPs and FRs; (iii) permute the composite matrix by exchanging rows and columns, so that all dominant entries appear on the main diagonal and get the derived DSM. A simple example in Fig. 1.4 can illustrate such a conversion process.

When the DP's DSM has been derived, the physical structure could be built with the synthesis of DPs in preparation for concept selection. This step can be modeled mathematically as

$$\{DP\} = [S]\{SC\} \tag{1.9}$$

where SC is a structure component array and $[S]$ is the synthesis matrix formed from the binary variable S_{ik}. S_{ik} equals 1 if DP_i is part of the component SC_k and 0 otherwise. Therefore, based on the DP's DSM, the SC's DSM could be further obtained representing the relationship among the components.

Once a product description is given, its properties, including those related to product property (PP) can be derived from the related property knowledge. Denoting the property knowledge k^p and the property knowledge domain Γ, we have

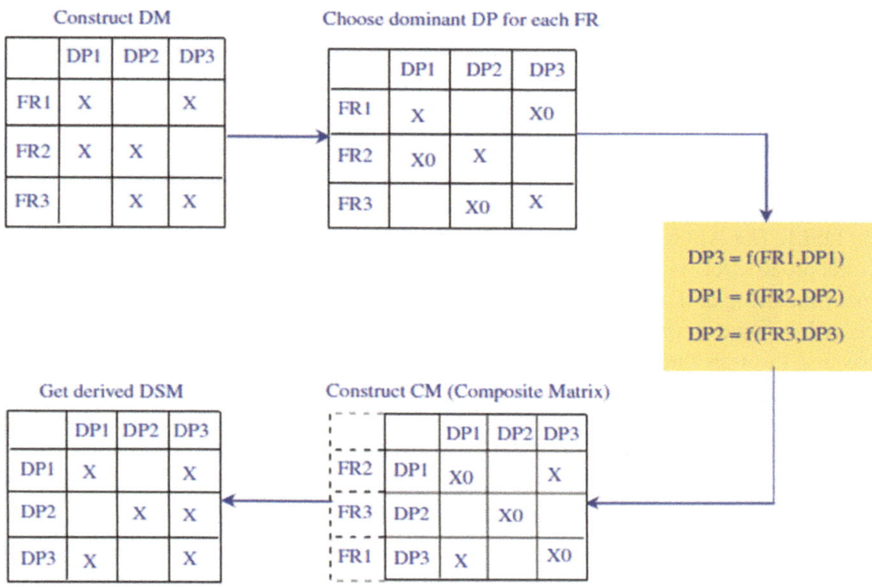

Fig. 1.4 Conversion of DM to DSM

$$\exists k^p \subseteq \Gamma : DSM \rightarrow PP \tag{1.10}$$

With the derived DSM and predicted PP, the product structure solution can be evaluated whether it can meet (i.e. is close enough to) the DC such as component compatibility, manufacturability, cost, safety, etc., which can be described as below:

$$\lambda_{\text{conflict}}(PP, DC) \Rightarrow \neg\lambda_{\text{specify}}(FR, CR) \vee \neg\lambda_{\text{satisfy}}(SP, FR) \tag{1.11}$$

where $\lambda_{\text{conflict}}$ means that the predicted PP has conflicts with DC, and the symbol '$\neg$' means negative. If there are any conflicts, the previous design solution needs to be improved or refined. It is necessary to distinguish the FRs from the DC. FRs would be those statements demanding the explicit presence of a particular feature, whereas DC is typically the conditions that must not be explicitly violated by a design solution.

1.3.2 Recursive Interaction of DM and DSM

When the design problem is initially set, the goals are usually vague, and many constraints and criteria are unknown. The function formulation, therefore, may embody inconsistencies emerging in the process of problem-solving. Furthermore, formulation of the function is solution-dependent, and the way the solution is conceived influences the way the function is conceived [17]. Decomposing high-level complex functions into lower level ones has been a common practice to create design concepts, and the AD suggests the zigzag mapping process to perform the decomposition of FRs and identification of suitable DPs. It has been noted that AD cannot deal with the design technical feasibility, which can be addressed based on DSM. The AD with DM is actually a conceptual-level functional model. When a conceptual-level model is embodied into a manufacturable entity, issues of form, manufacturing and assembly arise [18]. The conceptual-level functional model only describes what a product does when acting on material, energy or signal flows passing through the system boundary. Conceptual-level models do not relate how the solutions to artifacts are actually pieced together. For example, for a coffee grinder product, the function of ‘convert electrical energy into mechanical energy’ would most easily be solved by using a standard electric motor. What is not shown by the conceptual-level functional model is how the motor is physically connected to the other parts chosen to perform their need-based function. There exist an almost infinite number of choices of how the motor would be secured within the object. All of these attributes are not defined by the axiomatic functional design model and remain choices of the designer to configure the overall form of the product. In this research, DSM focusing the design technical feasibility can bridge this gap in representation. The DSM-based technical feasibility includes (i) design modularity, (ii) assembly and manufacturing feasibility, and (iii) cost and other constraints imposed. Through integrating DSM into AD process, the result of DSM-based technical feasibility check can give the support for DP selection, or feedback to improve the design mapping. The DSM-based resulting design may affect the functionality of the systems, for example, grouping DPs or integrating components together may couple functions. If that is the case, FRs may need to change or more DPs may need to be added.

Following the zigzagging process of AD, the integration of AD and DSM is proposed to be performed though a recursive interaction of DM and DSM (see Fig. 1.5). Each recursive step includes the following actions:

(i) The FR^{I} at a given level is mapped to the same level DP^{I}, and the responding DM^{I} is constructed under the guidance of selected solution principle, namely $SP \sim DM^{I}$.
(ii) The DM^{I} is conversed into related DSM^{I} through the matrix permutation method introduced before, and the design synthesis is conducted based on the derived DSM^{I}. The derived DSM^{I} would be evaluated through the design technical feasibility check and the predicted PP analysis.
(iii) If the evaluation of DSM^{I} is passed, then go to next level FR^{I+1}, and repeat the above actions. Otherwise, the FR^{I} with related SP shall be revised and get a new DM and DSM at the same level and repeat (i) and (ii).

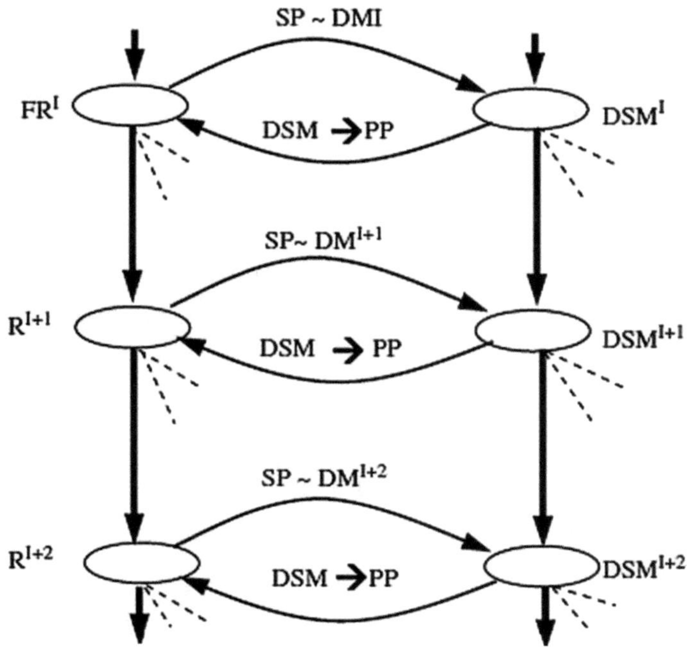

Fig. 1.5 Recursive interactions of DM and DSM

In conclusion, the recursive interaction of DM and DSM may result in the revision of FR or related SP. The reasoning steps and the sequence of FR/SP revision are described in Fig. 1.6. The first revision way is that the SP keeps the same, while the FR is refined or modified. The first revision way is that the SP keeps the same, while the FR is refined or modified. The FR revision has two sub-paths: (1) within a fixed function definition domain Ω, a FR is revised to a new one (namely FR*), (2) a FR is revised to a new one (namely FR*) within a new function definition domain Ω_{new}. These two FR revision steps are formulated in Eqs. (1.12) and (1.13) respectively.

$$\exists fr \in \Omega, FR^* = (FR \cup \{FR\}) : \lambda_{\text{specify}}(FR^*, CR) \tag{1.12}$$

$$\exists FR^* \in \Omega_{new} : \lambda_{\text{specify}}(FR^*, CR) \tag{1.13}$$

With a new function requirement FR* and the same SP, a recursive design mapping will be conducted. A new DM (namely DM*) is constructed and a new DSM (namely DSM*) is derived, which can be formulated as

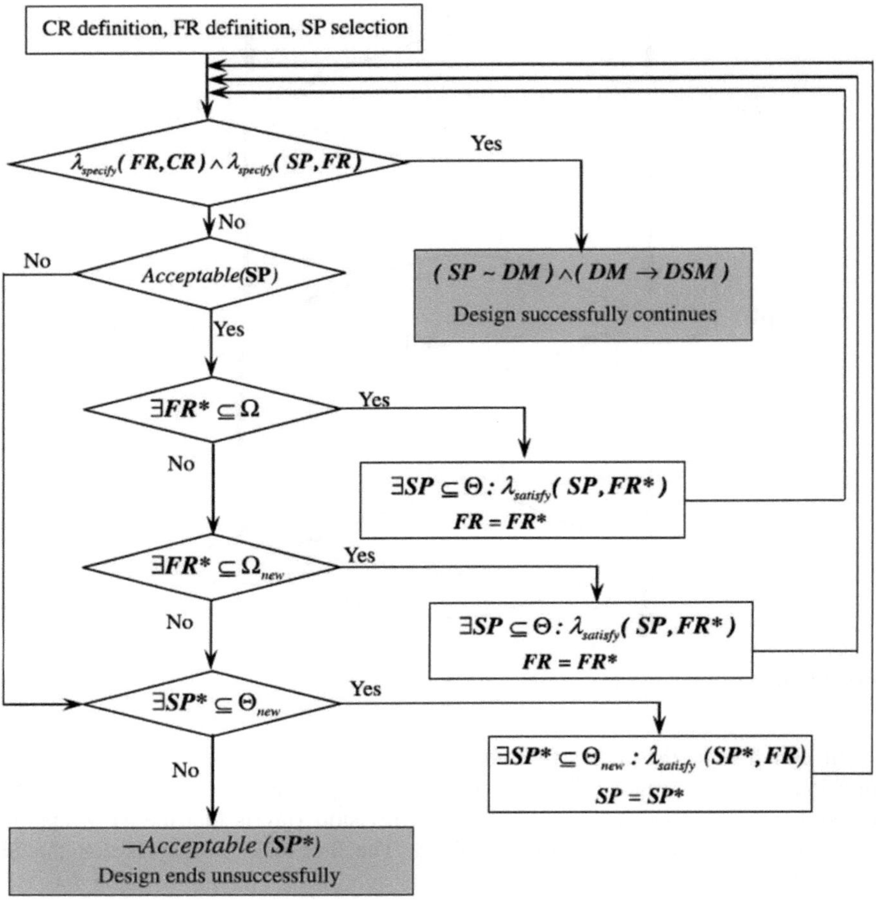

Fig. 1.6 Reasoning steps of FR/SP revision

$$(\exists SP \subseteq \Gamma : FR^* \rightarrow DP^*) \sim DM* \rightarrow DSM* \tag{1.14}$$

If the DSM* cannot meet the design technical feasibility, the designer will go to the second revision way, namely SP revision in a new domain theory Θ_{new}. For example, for a shock absorber solution, a designer may change the spring-based principle to the simple pneumatic piston-based principle. The SP revision way can be formulated as

$$\exists SP^* \in \Theta_{new} : \lambda_{satisfy}(SP^*, FR) \tag{1.15}$$

$$(SP^* : FR \rightarrow DP^*) \sim DM^* \rightarrow DSM^* \tag{1.16}$$

The recursive steps represent a designer's attempt to address a given design problem from both function and structure view of the product. During the recursive interaction process of DM and DSM, the FR redefinition occurs in terms of the previous function state, the contribution of the newly chosen solution (DP) towards solving the provided functions, and the additional design requirements the modified solution imposes. At each interaction level, various knowledge sources are consulted in order to take into consideration of constraints originating from all stakeholders. The knowledge sources include unstructured ones (e.g. employees' tacit knowledge) as well as structured/coded sources. Examples of the latter include DSMs of past designs (also processes).

1.4 System Development and Application

1.4.1 System Development

The framework of the AD–DSM integration design system is shown in Fig. 1.7. The top level components within the system are FR definition, DM construction, DSM deriving, DSM-based structural synthesis and design technical feasibility check. The DM construction is guided by the repository of solution principles, allowing the user to consult the FR–DP mapping knowledge.

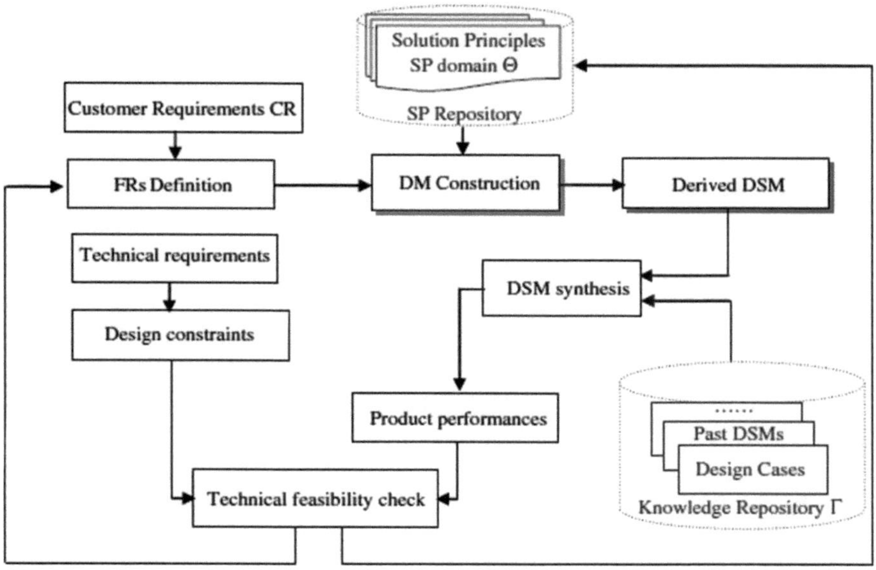

Fig. 1.7 Framework of the AD–DSM integration design system

The process of SP consulting for the FR–DP mapping is as follows: (1) designer inputs a desired function requirement in the form of verb attributes; (2) the design system searches the domain of solution principles for possible solutions according to the given verb and attributes; and (3) the design system outputs related solution principles, and designer could visit the introduction of each corresponding SP, which refers to a possible conceptual design solution. Figure 1.8 shows a screen shot of SP consulting for a FR–DP mapping. The AD suggests the zigzag process, while it is designer's job to perform the identification of sustainable DPs for specified FRs. Meanwhile, the designer can have the chance of selecting the optimum SP from a set of solution alternatives, which can potentially lead to the emergence of products of high-quality; and in redesign, designers are able to substitute better SPs for those old unsatisfactory ones. With the consulting aid of solution principles, the designer will be helped to identify most suitable means to carry the functions and construct the DM.

As stated before in this chapter, DM serves to define the manner in which the DPs will satisfy the FRs. However, the effects of decisions relating to the product such as cost, capacity, and physical integration, are not dealt with particularly well. The architecture of the product is defined not only by the decomposition of the complete product into elemental DPs or components, but also by the interactions between these DPs or components. The interactions may include well-specified interfaces and undesired or incidental interactions. As a result, a DSM needs to be generated to accommodate these issues and find undesired interactions. Therefore,

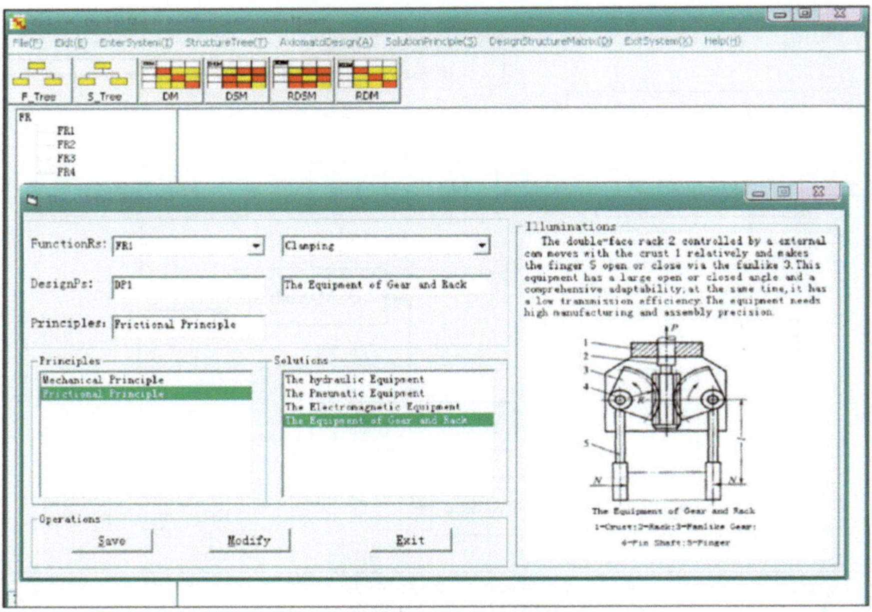

Fig. 1.8 Solution principle consulting for a FR–DP mapping

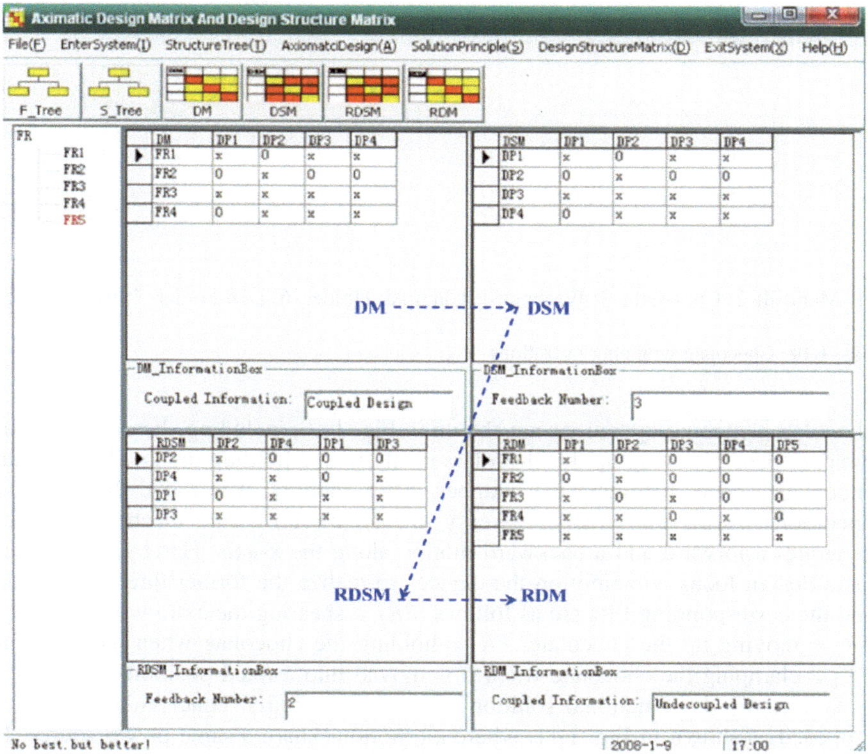

Fig. 1.9 Screenshot of DM–DSM conversions

the conversation of a DM to a DSM shall be conducted. The derived DSM may need to be re-engineered (so-called RDSM) to find feedback loops of interactions which could reveal the physical or capacity problems. With the result of RDSM, the original DM may be backtracked and modified to a refined one (called RDM). The developed system can support a designer to perform such a recursive process (see Fig. 1.9). It is noted that a product has various kinds of DSM according to different domains such as physical connection, material, cost, layout, etc. The derived DSM from DM at the conceptual design phase is not totally equal to real final DSM of designed product, and it could act as a base to deduce the final real DSMs of different domains.

1.4.2 Application of Integration Between AD and DSM

In this section, a conceptual design example of chocolate wrapping mechanism is given to illustrate the power of integration between AD and DSM. The normal

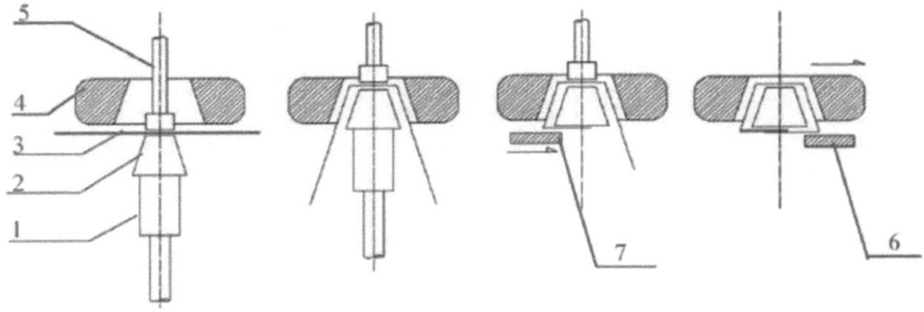

1: Mandrill 2: Chocolate 3: Wrapper 4: Clamp 5: Holder 6: Left Folder 7: Right Folder

Fig. 1.10 Chocolate wrapping operations

chocolate wrapping operations are shown in Fig. 1.10, including shearing the extra wrapper paper, moving up the chocolate firmly with a holding organ, clamping the chocolate, and then folding the wrapper. After analysis, it is found that the final operation (folding the wrapper) is easy to realize with a cam mechanism which generates a forward and a backward motion along the x-axis. Hereby, the mechanism design focus is mainly on the devices to realize the former three operations, and the corresponding FRs are as follows, FR_1 = shearing the extra wrapper paper, FR_2 = moving up the chocolate, FR_3 = holding the chocolate when moving, and FR_4 = clamping the chocolate when it is moved into a fixed position.

Guided by the functional solution principles, an initial conceptual solution is proposed as shown in Fig. 1.11, where all motions share a same power source. In the initial solution, the DPs are defined as follows; DP_1 (the design organ to shear the extra wrapper paper), which is realized by a combination of a cam-follower-spring; DP_2 (the design organ to move up the chocolate along the y-axis), which is realized by a cam-rocker; DP_3 (the design organ to hold the moving chocolate along the y-axis), which is realized by a combination of bevel gear-worm, cam-follower, and a spring; and DP_4 (the design organ to clamp the chocolate), which is realized by a clamp. The DM of the initial conceptual solution at the overall level is described as:

$$\begin{Bmatrix} FR_1 \\ FR_2 \\ FR_3 \\ FR_4 \\ FR_5 \end{Bmatrix} = \begin{bmatrix} X & 0 & 0 & 0 & 0 \\ X & X & 0 & 0 & 0 \\ X & X & X & 0 & 0 \\ X & X & X & X & 0 \\ X & X & X & X & X \end{bmatrix} \begin{Bmatrix} DP_1 \\ DP_2 \\ DP_3 \\ DP_4 \\ DP_5 \end{Bmatrix} \tag{1.17}$$

Obviously, the DM at the overall level is decoupled, and the initial conceptual solution seems feasible. After deep investigations, the FR_3 can be decomposed into four sub-function requirements, namely, FR_{31} (providing power supply), FR_{32} (transferring the power for kinematical motion), FR_{33} (motion transformations to move the holder downwards along y-axis), and FR_{34} (upturning the holder along

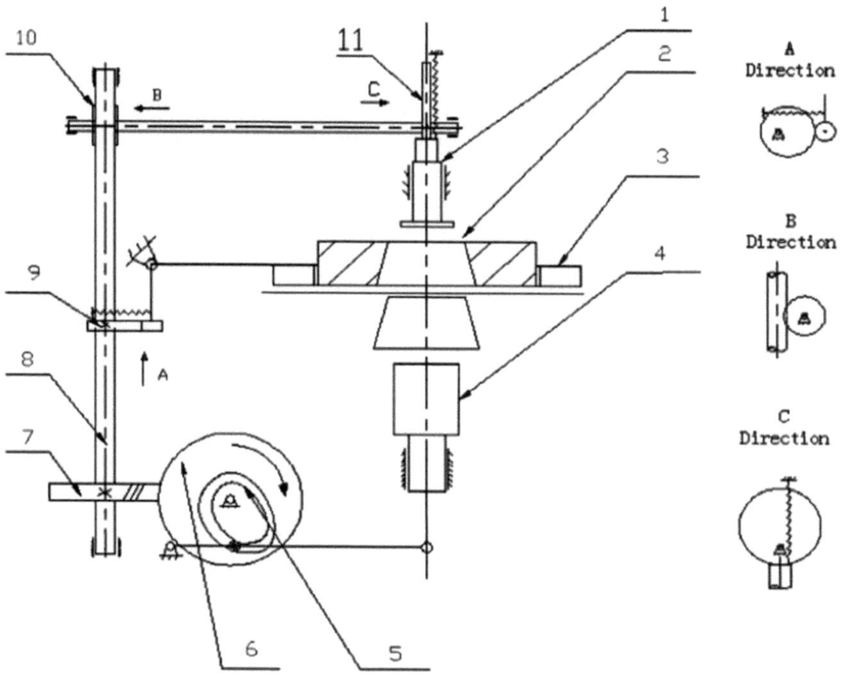

1: Holder 2: Clamp 3: Shears 4: Mandrill 5: Cam-Rocker 6: Active Gear 7: Driven Gear 8: Transmission Shaft 9: Cam-Follower-Spring 10: Worm-Gear 11: Cam-Follower-Spring

Fig. 1.11 The initial conceptual solution

the y-axis). The corresponding DPs are DP_{31} (the power of the motor), DP_{32} (bevel gear-worm), DP_{33} (cam-follower), DP_{34} (coefficient K of the spring). And the design matrix $[DM_3]$ to decompose the FR_3 mapping is

$$\begin{Bmatrix} FR_{31} \\ FR_{32} \\ FR_{33} \\ FR_{34} \end{Bmatrix} = \begin{bmatrix} X & 0 & 0 & 0 \\ X & X & X & 0 \\ X & X & X & X \\ 0 & 0 & X & X \end{bmatrix} \begin{Bmatrix} DP_{31} \\ DP_{32} \\ DP_{33} \\ DP_{34} \end{Bmatrix} \tag{1.18}$$

After choosing diagonal elements of $[DM_3]$ as output variable and the related [DSM] for the third design organ (DP_3) can be derived as

$$[DSM_3] = \begin{matrix} DP_{31} \\ DP_{32} \\ DP_{33} \\ DP_{34} \end{matrix} \begin{bmatrix} X & 0 & 0 & 0 \\ X & X & X & 0 \\ X & X & X & X \\ 0 & 0 & X & X \end{bmatrix} \tag{1.19}$$

The derived DSM in Eq. (1.19) shows feedback loops between DP_{32}–DP_{33} and DP_{33}–DP_{34}. However, these feedback loops had not been anticipated in the design. Analysis of the derived DSM revealed that the likely reason for such appearance was a structural redundancy issue. The structural redundancy actually always comes from functional redundancy. Is it possible to simplify the motion transformation mechanism and get rid of the spring? With this question and after a refinement, the second conceptual solution is proposed in Fig. 1.12. The FRs for the refined solution are FR_{31} (providing power supply), FR_{32} (transferring the power for kinematical motion) and FR'_{33} (motion transformations to move the holder downwards and upwards). The corresponding DPs are DP_{31} (the power of the motor), DP_{32} (the helicoids groove of the cylindrical cam), DP_{33} (the length of the crank). The refined design matrix $[DM'_3]$ is as follows:

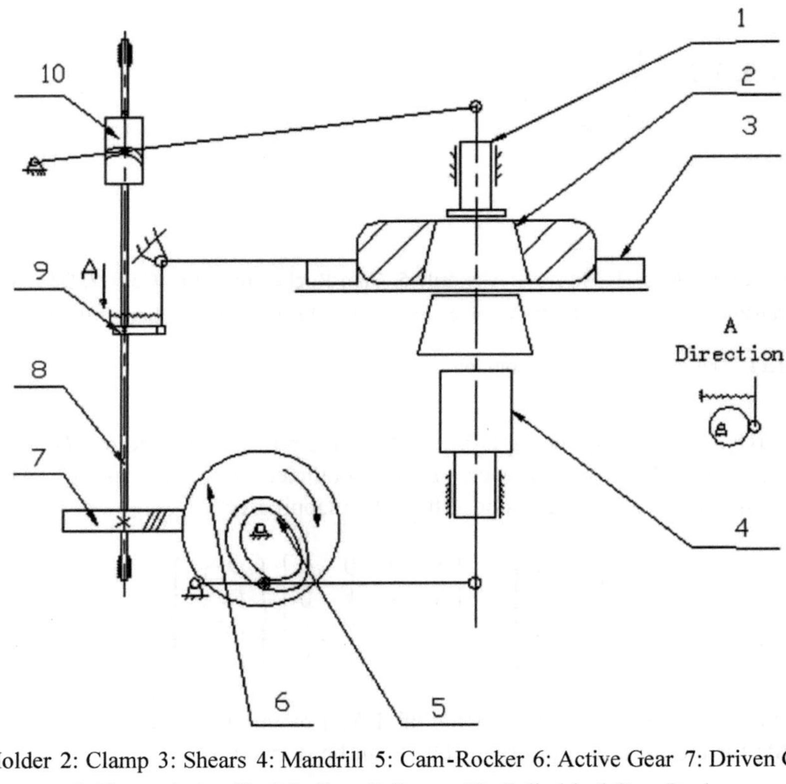

1: Holder 2: Clamp 3: Shears 4: Mandrill 5: Cam-Rocker 6: Active Gear 7: Driven Gear 8: Transmission Shaft 9: Cam-Follower 10: Cylindrical Cam-Rocker

Fig. 1.12 The refined conceptual solution

$$\begin{Bmatrix} FR_{31} \\ FR_{32} \\ FR'_{33} \end{Bmatrix} = \begin{bmatrix} X & 0 & 0 \\ X & X & 0 \\ 0 & X & X \end{bmatrix} \begin{Bmatrix} DP'_{31} \\ DP'_{32} \\ DP'_{33} \end{Bmatrix} \tag{1.20}$$

In the same way of DM–DSM conversion, the related DSM for the third modified design organ can be derived as

$$[DSM'_3] = \begin{matrix} DP'_{31} \\ DP'_{32} \\ DP'_{33} \end{matrix} \begin{bmatrix} X & 0 & 0 \\ X & X & 0 \\ 0 & X & X \end{bmatrix} \tag{1.21}$$

In $[DSM_3]$ shown in Eq. (1.19), there are two feedback loops between DPs. In $[DSM'_3]$ shown in Eq. (1.21), the number of feedback loops reduces to zero, which is in favor of minimizing rework and minimum cost of change. Meanwhile, the refined design in Fig. 1.12 has less number of kinematics building blocks than the initial design in Fig. 1.11, which means cheaper product cost and more convenient maintenance. In short, the refined solution has resolved the structure redundancy issue, and simplified the structure design, thus arriving at better-quality conceptual design.

1.5 Conclusion

Conceptual design problems are usually complex and involve various functional requirements and a large number of potential means as solutions. AD suggests the zigzag mapping process to decompose the function requirement and provides two axioms to judge the quality of the function decomposing. Like many design methods, AD has been applied in a variety of areas, as researchers and practitioners have sought to leverage its advantages. Along the way, however, its practicability has been challenged. Under the AD context, the designer is not able to master the interactions amongst the DPs, including geometry, spatial layout, interfaces (e.g. logical and physical connectivity), which will decide the quality of structural design. DSM has been proved that it can enhance AD on such issue. Therefore, this chapter has investigated the logic of integration between AD and DSM, and proposed a recursive interaction mechanism between AD's DM and DSM to realize such integration. A computer-aided conceptual design system has been developed to realize the proposed integration model of AD and DSM. Further practical use cases to evaluate and validate the proposed idea and the developed prototype form the scope of our future work.

References

1. Suh NP (1990) The principles of design. Oxford University Press, Oxford
2. Suh NP (2001) Axiomatic design: advances and applications. Oxford University Press, Oxford
3. Dong Q, Whitney D (2001) Designing a requirement-driven product development process. In: Proceedings of the 13th international conference on design theory and methodology (DTM 2001), 9–12 Sept 2001, Pittsburgh, Pennsylvania, USA
4. Guenov MD, Barker S (2004) Requirements-driven design decomposition: a method for exploring complex system architecture. In: Proceedings of ASME 2004 international design engineering technical conferences and the computers and information in engineering conference (DETC'04), Sept 28–Oct 2, 2004 Salt Lake City, USA
5. Steward D (1981) The design structure system: a method for modeling the design of complex systems. IEEE Trans Eng Manag 3:71–74
6. Eppinger SD (1991) Model-based approaches to managing concurrent engineering. J Eng Des 2:283–290
7. Eppinger SD, Whitney D, Smith R et al (1994) A model-based method for organizing tasks in product development. Res Eng Des 6:1–13
8. Eppinger SD, Salminen V (2001) Patterns of product development interactions. In: International conference on engineering design, ICED 01, Glasgow, 21–23 Aug 2001
9. Tang D, Zheng L, Li Z et al (2000) Re-engineering the design process for concurrent engineering. Comput Ind Eng 38:479–491
10. Yassine A, Falkenburg D (1999) Design process specifications management. J Eng Des 10:223–234
11. Yassine A, Whitney D, Eppinger SD et al (2003) Connectivity maps: modeling and analysing relationships in product development processes. J Eng Des 14(3):377–394
12. Suh NP, Sekimoto S (1990) Design of thinking design machine. Ann CIRP 39:145–148
13. Park KJ, Kang BS, Song KN, Park GJ (2003) Design of a spacer grid using axiomatic design. J Nucl Sci Technol 40(12):989–997
14. Chin WS, Kwon JW, Lee DG (2003) Trenchless repairing of underground pipes using RTM based on the axiomatic design method. J Compos Mater 37(12):1109–1126
15. Chen KZ, Feng XA, Zhang BB (2003) Development of computer-aided quotation system for manufacturing enterprises using axiomatic design. Int J Prod Res 41(1):171–191
16. El-Haik BS (2005) Axiomatic quality: integrating axiomatic design and six-sigma, reliability, and quality engineering. Wiley-Interscience, Hoboken
17. Cross N (2000) Engineering design methods: strategies for product design, 3rd edn. Wiley, New York
18. Bohm MR, Stone RB (2004) Representing functionality to support reuse conceptual and supporting functions. In: Proceedings of ASME 2004 international design engineering technical conferences and the computers and information in engineering conference (DETC'04), Sept 28–Oct 2, 2004 Salt Lake City, USA

Chapter 2
Product Design Knowledge Management Based on Design Structure Matrix

2.1 Introduction and Synopsis

Knowledge Management (KM) is currently receiving increasing attention in diverse areas. The concept of KM can be defined in the way to organize information or data. It is possible to think in the evolution from data to knowledge in the following way: data → information → knowledge [1]. It is a common sense that the past experience and information may play a crucial role in a new product development process. The main objective of KM could be understood to capture and store all the past experience and information, and make them usable in assisting the development of new products or in the enhancement of existing ones. On one hand, knowledge can be seen as a representation of the real world, on the other, it can be conceptualized as a product of the interaction between individual cognition and reality [2]. Ensuring efficient knowledge capture, sharing and reuse to support new product development, it is important to develop a structured way of processing the past experience and information behind the elements interactions.

As a structured modeling method, more recently the design structure matrix (DSM) model has been regarded as a good roadmap of system knowledge [3]. The process of making DSMs involves understanding and capturing the relation between system elements, which is a process of knowledge capture. DSM therefore is an excellent delivery medium of knowledge and as well as an arena for knowledge creation and management. Comprehensive research and development requirements along with a cogent framework, however, have not been proposed for integrating KM and DSM. This chapter tries to consolidate the KM method based on DSM and identify research and development issues in a unifying manner.

The remainder of the chapter is organized as follows. Section 2.2 reviews related research results. Section 2.3 proposes a comprehensive framework to illustrate KM lifecycle based on DSM. Section 2.4 focuses on how to create and capture both

D. Tang et al., *Matrix-Based Product Design and Change Management*,
DOI 10.1007/978-981-10-5077-0_2

direct and indirect knowledge from different kinds of DSMs. Section 2.5 introduces how to organize and use the captured knowledge from DSMs. Section 2.6 concludes the chapter.

2.2 Related Research of Knowledge Management

It is widely observed that the society we live in has been gradually turning into a "knowledge society", and knowledge management has now been a critical success issue. It has been discovered that inventive problems in completely different areas have been solved with similar principles. This assists in stating why is so important to maintain a well-organized knowledge. In any development process, problems arisen, past experience and information may play a crucial role, to help find a solution in less time. In reality, the product related knowledge is distributed across the whole company, and this kind of know how represents an essential resource for successful competition in the market and should therefore be preserved and used as efficiently as possible. The pressure to shorten the time to market for new products and therefore to shorten development time requires an adequate management of knowledge possessed by a company. Normally, a knowledge management effort includes knowledge eliciting and recording. The resulting knowledge storage are intended to be live, up to data documents synthesizing the corporate expertise in the form of standards, rules, procedures, and best practices. In addition, knowledge management enables the recorded expertise becoming more easily reviewed, shared, and used [4–6].

Knowledge management can be conducted at different levels such as part or individual level, assembly level, and system level. Based on the interview at industries, Dong finds a trend that as design scale gets bigger, less knowledge is documented [3]. In other words, the knowledge capturing and recording is more poorly implemented at the system level than part or individual level, especially for the large-scale complex system. With part or individual knowledge, each person sees only part of the entire system, which could cause inconsistent understanding problems in project development. Although ideas are formed in the minds of individuals, interaction between individuals typically plays a critical role in developing these ideas. That is, communities of interaction would contribute to the amplification and development new knowledge.

System level knowledge can be defined as the knowledge concerning system interactions. Knowing system interactions in the product development process is critical for project management and decisions. System interactions mainly refer to the inter-relationships existing between element types and between element types during the product development cycle. Capturing inter-relationships between affiliated system elements is necessary if the product engineering is to be handled (i.e. analyzed, engineered, documented, developed, and managed) efficiently and effectively. These elements can be many things: physical components that make up the product, performance attributes expected by the consumer, engineering requirements the product must meet, or tasks that create the product. Each of these different element types contributes to the development of a product. The idea of using DSM to represent the system interaction is not new. DSM method was proposed for system modeling and analysis more than 30 years ago. Steward firstly introduced DSM and developed some algorithms for manipulating the matrix as tools for systems design and analysis [7, 8]. However, it is not until recently that DSM methods started to attract attention for managing the complexity of large engineering systems and complex product development processes [9–13].

Most of the aforementioned DSM-based researches provide some insight into the interaction structure of a complex system or process (i.e. a dependency map); however, they fail to expose and explore the logic behind these dependencies. That is, if an element B is shown to interact with another element C, it is not clear why and how this relation exists and where it stems from [13]. Meanwhile, most of current DSM research is focused on the partitioning algorithms for system re-engineering, while the knowledge behind the DSM re-engineering is worthy of paying attention.

To conclude, system level knowledge is critical to the success of the design of large-scale systems, yet is often missing due to its empirical nature. The knowledge concerning system interactions requires the identification and understanding of the inter-relationships among the different elements. DSM, a structured method, has been proved to be most suitable for not only capturing system level interactions, but also providing analysis to the system interactions to aid the design and management of the system. However, how to deal with the system knowledge management based on DSM still deserves extensive research.

2.3 KM Lifecycle Framework Based on DSM

According to the KM lifecycle, a DSM-based knowledge management framework is proposed in Fig. 2.1. As an input of the knowledge management lifecycle, the system and its scale need to be defined at first. Following tasks are suggested to finish the KM lifecycle:

1. Knowledge acquisition

Identify and capture the knowledge used in the design process for solution, documentation and storing.

- *System interaction study*

List all system elements and study the information flow between them. The information flow can be got from the design document and interview with experienced designers.

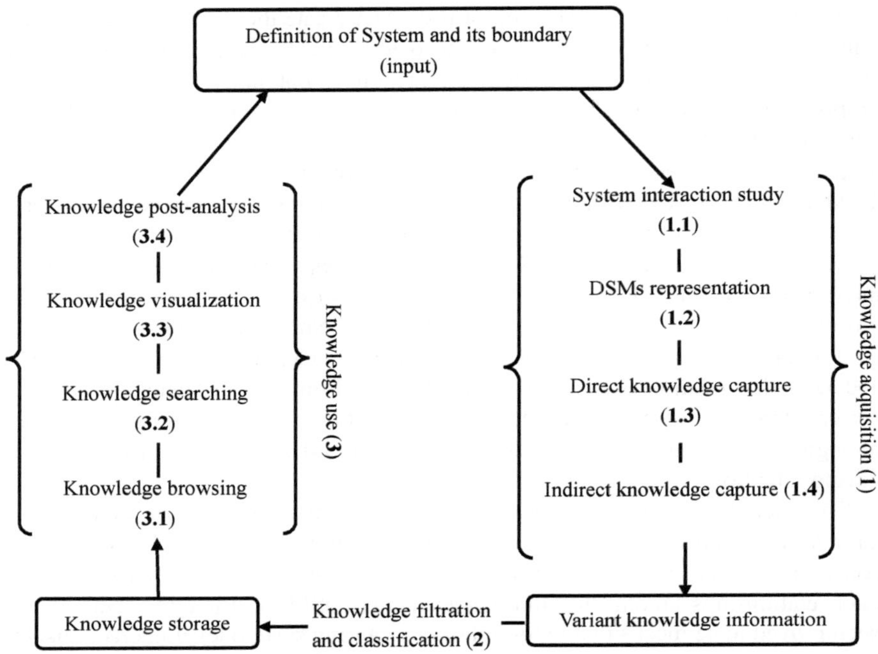

Fig. 2.1 DSM based KM lifecycle framework

- *DSMs representation*

Build binary matrices to represent the information flow between system elements in product, process and organization domains.

- *Direct knowledge capture*

Focus on the direct interactions between system elements, extract the know-how behind them.

- *Indirect knowledge capture*

Focus on the indirect interactions among system elements, analyze the interaction source and trace the know-how behind them.

2. Knowledge organization

Filter the valuable knowledge and classify type of knowledge. This is done to support efficient storing and reuse of the formalized knowledge (algorithms, rules, relations and data).

3. Knowledge use

By retracing the design history knowledge, the designer is able to find the design decision, rational and related information that govern the existing product and from where new design could originate.

- *Knowledge browsing*

  Using DSM as a knowledge browser, the designer could locate his interested information conventionally.

- *Knowledge searching*

  Using some key words, the designer could obtain the related knowledge rapidly.

- *Knowledge visualization*

  Use graphic model to transparently represent the result of knowledge capturing.

- *Knowledge post-analysis*

  Account the reuse frequency and clarify the quality of captured knowledge. Evolve and consummate the system knowledge based on the evaluation of designer.

2.4 Knowledge Acquisition

2.4.1 Direct Knowledge Capture from DSMs

Development of complex products and large systems is a highly interactive social process involving hundreds of people designing thousands of interrelated components and making millions of coupled decisions. Generally, the complexity of a large system can be modeled through three domains of product development: a product domain, process domain, and an organization domain, which are explained as follows [14].

1. Product: A complex product or large system is decomposed into sub-systems, and these in turn may be further decomposed into sub-assemblies and/or components.
2. Process: A full development process is decomposed into phases or sub-processes, and these in turn may be further decomposed into tasks or activities.
3. Organization: A large development organization is decomposed into teams, and these in turn may be further decomposed into working groups and individual actors.

It is able to learn about the complex social phenomenon of product development by studying the patterns of interaction across the decomposed elements within each domain [14]. DSM is always regarded as the suitable tool to describe the interactions and document the decomposition of each of the three domains.

To construct a DSM, one assigns the individual system elements of a domain to the rows and columns of a square matrix. Then, going down the list, if element *b* is an input to element *c*, one puts a mark in the column of *b* and the row of *c*. One continues until all (known) hierarchical and interdependent relationships have been accounted for. Marks below the diagonal represent forward information transfer which affects the later element. Marks above the diagonal represent information fed back to earlier element (see Fig. 2.2).

It is not enough to know only the form of DSM, it is also necessary to know the meaning of its marks and the knowledge inside. In a DSM, the off-diagonal marks are not simply placeholders. Corresponding to each is a specific real issue, involving at least two system elements. Thus the matrix understates the depth of the designers' knowledge for them; each of the marks stands for a specific problem or question.

Things are traceable, if they leave traces. The system level knowledge can be traced through the element interactions in DSM which can be regarded as a knowledge map. In other words, the system knowledge can be depicted by the marks in a DSM, and interpreted by the syntactic and semantic information of the element interactions.

A single-domain DSM can only reflect the system knowledge in an isolated area. For example, through a product DSM, we can directly know the physical and logical interdependencies among product elements (such as components). To capture the system level knowledge from a comprehensive view, we need to analyze the combined DSMs of three domains defined above. Generally, it can be assumed that product DSM, process DSM, and organization DSM are isomorphic to each other. That is, if component *j* either imposes geometry constraints or transfers forces, material, energy, and/or signals to component *i*, for component *i* to function

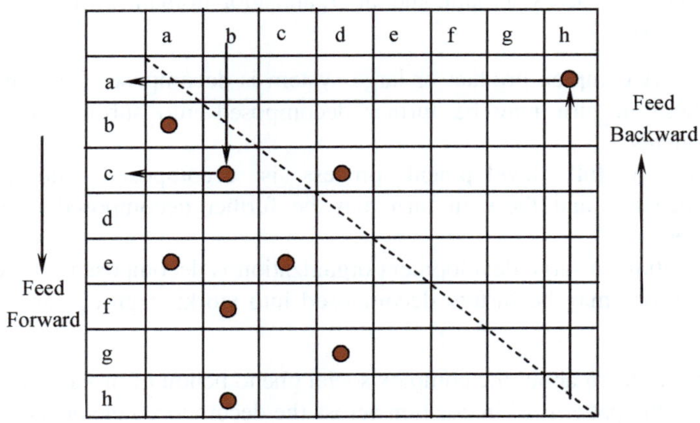

Fig. 2.2 DSM with feed forward information and feedback information

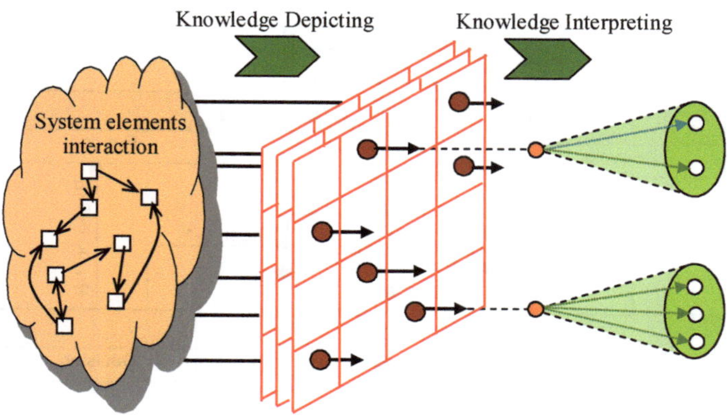

Fig. 2.3 Direct knowledge capture from isomorphic three-domain DSMs

properly, technical information will flow from team *j* (activity *j*) to team *i* (activity *i*). Therefore, linking the same-position marks in the isomorphic DSMs (see Fig. 2.3), the comprehensive system knowledge can be depicted and interpreted by answering following questions:

- **What** are the product elements—design parameters, components, design representation objects, etc.—that are managed during product design and development, what are their characteristics and what are the links between them?
- **Who** are the actors or teams that play different roles in the creation, maintenance and use of those product elements and links?
- **Where** are the product elements represented: in physical media such as documents, drawings, spreadsheets or other files, or in less tangible things such as telephone calls, discussions, emails, meetings or people's brains?
- **How** can the product elements realized by different tasks or activities?
- **Why** are the product elements created, modified or evolved?
- **When** (in which lifecycle) are the product elements created, modified or evolved?

Figure 2.4 further illustrates the syntactic and semantic meaning of direct captured knowledge from three-domain DSMs which are isomorphic.

Under the isomorphic context, it is easy to understand that if two components have interactions, the teams (tasks) that design them also need to interact. However, in the development of highly complex products, it would be naive to expect a perfect mapping between three-domain DSMs. That is, due to product ambiguity and complexity, defined as the absence of knowledge about product variables and/or their interfaces, some product design interfaces or interactions are not foreseen at the outset of the project and are only discovered after design teams' work on the systems themselves [14, 15]. In other words, most of the product architectural knowledge is explicit and known by development organizations, while some interfaces or interactions between components are unspecified (or even

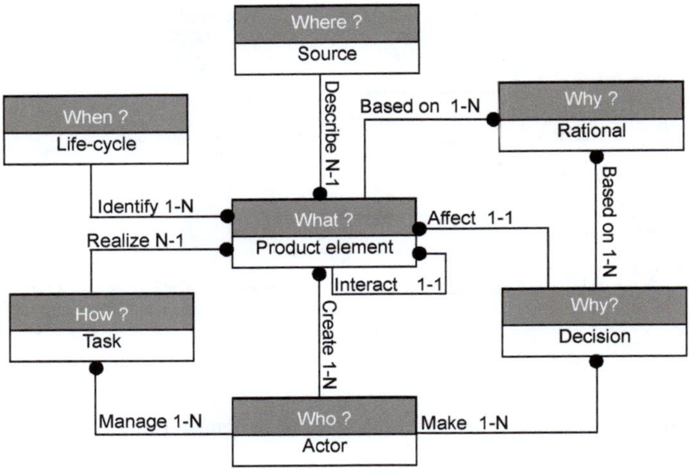

Fig. 2.4 Syntactic and semantic meaning of direct captured knowledge from isomorphic three-domain DSMs

unknown) and only identified or documented during the design process itself. It then becomes important to determine where (in the product) those unidentified interfaces or interactions are likely to be, and how they can be uncovered. Therefore, this kind of absent knowledge could be captured through the misaligned mapping between three-domain DSMs (see Fig. 2.5).

Based on the captured knowledge above, we could discover the design history of every feature of a product, and make the past product development traceable. The

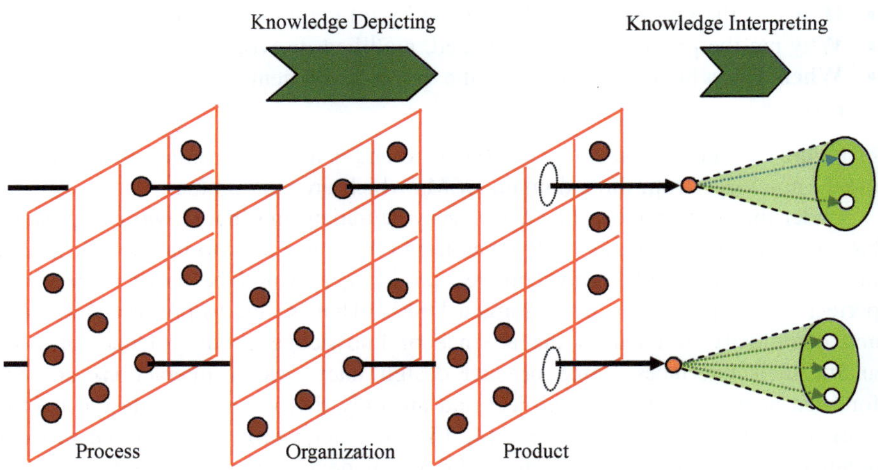

Fig. 2.5 Knowledge capture through misalignment between three-domain DSMs

knowledge can be referred to the traceability defined in [16], and enable tracing the product development activities, design routes, designers' actions, decisions, reasoning, events, etc. The three-domain DSMs could be knowledge browsers to navigate the evolution of design items, in both a forward and backward direction, i.e. from its origins, through its development and specification, to its deployment and realization, and through periods of on-going refinement and iteration in any of these phases. The DSM-based knowledge also can help predict the impacts of design changes through tracing designed items back to their sources. In addition, the captured knowledge can improve understanding of the design routes by linking designed items to rationales, decisions and assumptions behind them. Meanwhile, the knowledge assists on simulating design episode in another situation: for performing changes on existing solutions, reusing of the existing solutions in new projects, configuration of the new variant of the product, and educational process for inexperienced designers.

2.4.2 Indirect Knowledge Capture

In the above section, DSMs act as knowledge savers through analyzing direct interactions between system elements. However, just as shoals can exist hidden in the ocean, indirect interdependencies can exist [17]. As shown in a DSM in Fig. 2.6, there is a direct interaction from component *a* to component *b*, while there are two indirect component interactions "$a \rightarrow d \rightarrow b$" and "$a \rightarrow f \rightarrow b$". Meanwhile, in some instances, we use the concept of indirect team interaction to hypothesize that team *a*, whose component has a design interface with component *b*, may not report direct interaction with team *b* because it interacts with an intermediary team (team *k*, which also interacts with team *j*) which passes the information (to team *a*) that would otherwise have flowed directly from team *b* to team *a*. Such kind of indirect interdependencies or interactions can be fatal or merely annoying, require extra work. Or they can be fortunate, revealing new knowledge of design, which can be called indirect knowledge. Indirect interdependencies or interactions are one of the primary sources for uncertainty in design processes.

In a DSM, each of the off-diagonal marks may require actions to deal with the related questions or problems. In many instances, there would be tradeoffs and conflicts, some of which would require a costly and time-consuming dispute

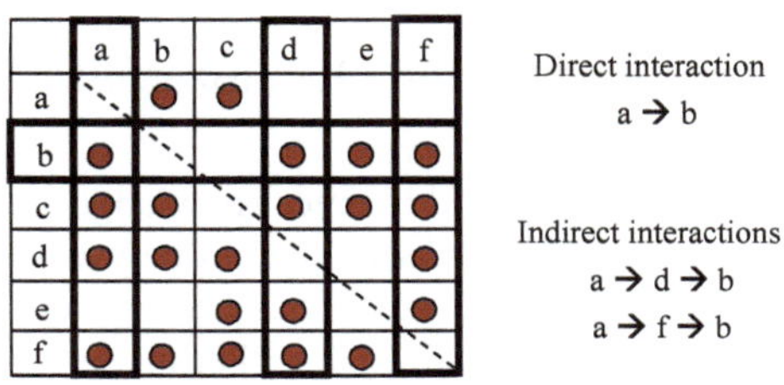

Fig. 2.6 Direct and indirect interaction in DSM (modified based on [18])

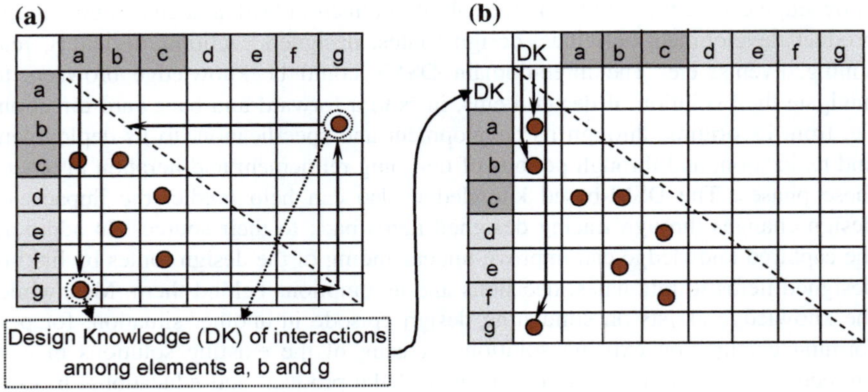

Fig. 2.7 **a** As-is DSM, **b** rationalized DSM with design knowledge

resolution process [17]. Especially the marks above the diagonal could cause process iteration or natively affect the modular product architecture. Researchers have invented various partitioning algorithms to re-engineer the DSM and eliminate the marks above the diagonal. However, the algorithms can only provide context-free analysis of the DSM, what's the knowledge behind them is still unknown. The essential knowledge for re-engineering the DSM is to identify the iteration's cause and then eliminate it. To capture this kind of knowledge, it is required to study each iterative block separately which is caused by indirect interactions and identify reasons for iteration.

For example, a fragment of a cycle depicted in Fig. 2.7a shows the indirect interactions among elements *a*, *b*, and *g* which cause serious iteration problem. If they have experience with previous designs, the designers will know that the problem is certain to come up. They will also have knowledge about what was done in previous designs, and how the resulting artifacts performed. This knowledge gives the designers a set of preformed strategies to eliminate the iterations. In particular, many of the routes to failure will be known, and presumably do not need to be tried again. Therefore, a rationalized DSM can be set up shown in Fig. 2.7b with the design knowledge of interactions among elements *a*, *b* and *g*, which is useful for not only reducing unnecessary feedback or rework but also designing modular product architecture.

2.5 Knowledge Organization and Application

Through direct and indirect knowledge capture from DSMs, we can get various and unstructured knowledge such as personal experiences, situational decisions, learned facts by trial and error, etc. For efficient use and reuse of them, this knowledge needs to be organized by filtering, refining, classifying, and rearranging. In this

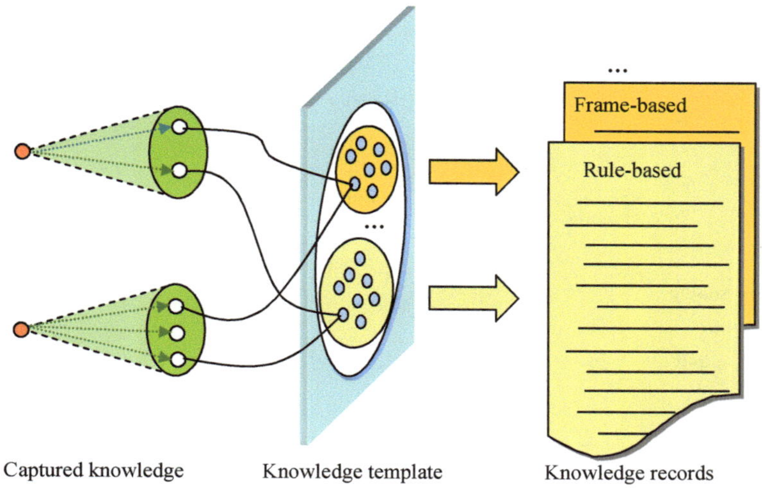

Fig. 2.8 Knowledge organizations by knowledge template

research, the knowledge organization is conducted by a knowledge template. The knowledge template can provide the structure and meta information of knowledge type to enable designers filtering valuable knowledge and avoiding knowledge overload, and to ensure that all useful knowledge (innovative information) is saved. The knowledge organization process is shown in Fig. 2.8.

The knowledge template is defined according to the most popular ways of knowledge representation. Therefore one can differentiate the captured knowledge into typical representations, such as *rule*-based (knowledge in terms of facts and rules for manipulating facts), *frame*-based (a frame, concept, schema or unit as a record like structure, a form for encoding on a stereotyped situation; associated with a frame is a set of attributes, the descriptions or values of which are contained in slots) and *semantic net*-based (semantic nets are similar to frames; it is a network or graph of nodes linked together by arcs, arcs represent relations) representations. After filtration and classification, the different types of knowledge records are stored with sequence number for future reuse. Although each type of knowledge has a distinct purpose, together they can support knowledge-intensive activities throughout the process lifecycle.

It is possible to store knowledge in all kinds of ways, but this does not make it real for people who will use it. In this research, a prototype has been developed to enable the knowledge browsing, searching and visualization. Taking a hairdryer product as an example, Fig. 2.9 shows a screen of knowledge browsing. The user can navigate the DSM inner cells to check the characteristics of the product architecture. Figure 2.10 gives another screen about the interaction knowledge searching through defining the level of interaction (strong, medium and weak), context of interaction (physical connectivity, cost, time, etc.) and the milestone of interaction (conceptual design, embodiment design, detailed design, process

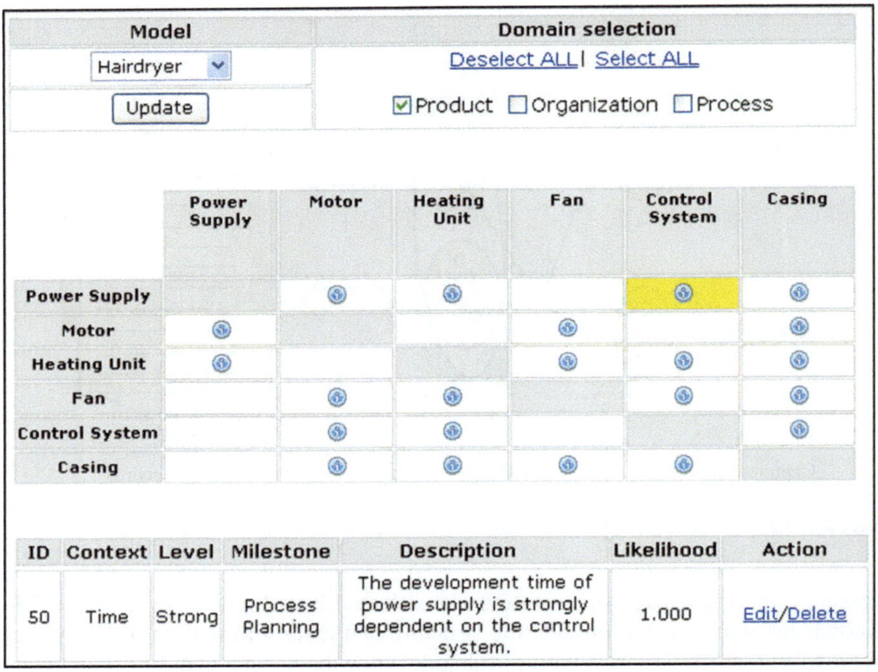

Fig. 2.9 Knowledge browsing

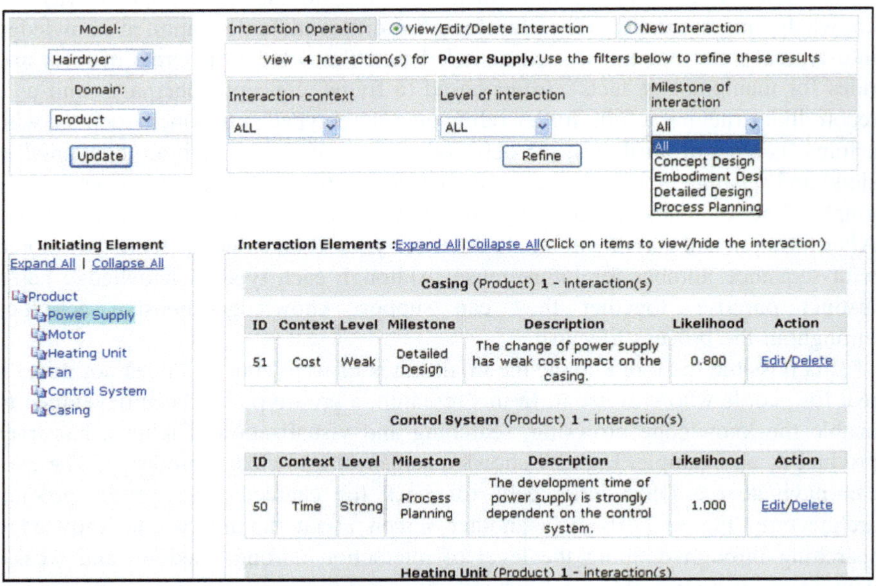

Fig. 2.10 Knowledge searching

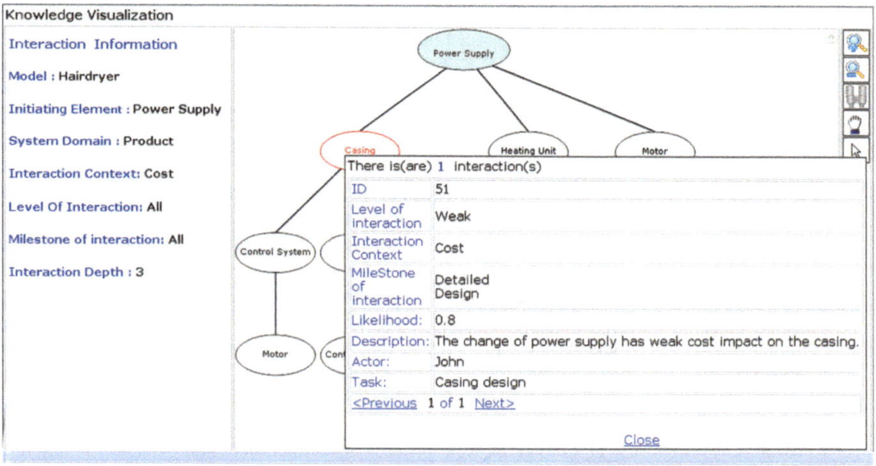

Fig. 2.11 Knowledge visualization

planning, etc.). To be able to perform a better display of system interaction knowledge, a knowledge visualizing function has been realized in the prototype (see Fig. 2.11). Through the visualization result, the designer can easily grasp the overall situation of the system knowledge, and could conduct the change propagation analysis for better decision making.

2.6 Conclusion

The system level knowledge management concerning system interactions is often missing due to its empirical nature. In this chapter, DSM is proposed to be suitable means to define, capture and organize and distribute the system level knowledge. How to capture the direct and indirect knowledge through DSMs is presented in detail.

Clearly more work remains to be done. The knowledge discussed in this chapter mainly refers to the explicit knowledge. How to capture the implicit knowledge through DSMs is still an open question. With the knowledge captured, how designers simulate and predict the impact of a design change is another future work.

References

1. Sorli M, Stokic D, Gorostiza A et al (2006) Managing product/process knowledge in the concurrent/simultaneous enterprise environment. Robot Comput Integr Manuf 22:399–408
2. Krogh GV (1998) Care in knowledge creation. Calif Manag Rev 40(3):133–153

3. Dong Q (2002) Predicting and managing system interactions at the early phase of product development process. PhD thesis, Massachusetts Institute of Technology
4. Mili F, Shen W, Martinez I et al (2001) Knowledge modeling for design decisions. Artif Intell Eng 15:153–164
5. Cheung WM, Bramall DG, Maropoulos PG et al (2006) Organizational knowledge encapsulation and re-use in collaborative product development. Int J Comput Integr Manuf 19(7):736–750
6. Jung J, Choi I, Song M (2007) An integration architecture for knowledge management systems and business process management systems. Comput Ind 58:21–34
7. Steward D (1962) On an approach to the analysis of the structure of large systems of equations. SIAM Rev 5:321–342
8. Steward D (1981) The design structure system: a method for modeling the design of complex systems. IEEE Trans Eng Manag 3:71–74
9. Eppinger SD (1991) Model-based approaches to managing concurrent engineering. J Eng Des 2:283–290
10. Eppinger SD, Whitney D, Smith R et al (1994) A model-based method for organizing tasks in product development. Res Eng Des 6:1–13
11. Yassine A, Falkenburg D (1999) Design process specifications management. J Eng Des 10:223–234
12. Tang D, Zheng L, Li Z et al (2000) Re-engineering the design process for concurrent engineering. Comput Ind Eng 38:479–491
13. Yassine A, Whitney D, Eppinger S et al (2003) Connectivity maps: modeling and analysing relationships in product development processes. J Eng Des 14(3):377–394
14. Eppinger SD, Salminen V (2001) Patterns of product development interactions. In: International conference on engineering design, ICED 01, Glasgow, August 21–23
15. Soe ME, Eppinger SD, Rowles CM (2004) The misalignment of product architecture and organizational structure in complex product development. Manag Sci 50(12):1674–1689
16. Štorga M (2004) Traceability in product development. In: International design conference—design 2004, Dubrovnik, May 18–21
17. Baldwin CY, Clark KB (2000) Design rules: the power of modularity, vol 1. The MIT Press, London
18. Clarkson PJ, Simons C, Eckert C (2004) Predicting change propagation in complex design. J Mech Des ASME Trans 126:788–797

Chapter 3
Matrix-Based Conceptual Solution Generation Approach of Multifunction Product

3.1 Introduction and Synopsis

Some products have one function and others have two or more functions. The former is a single-function product and the latter is a multifunction product. The multifunction product has become more prevalent in recent years, as customers desire both increased capabilities and reduced complexity to decrease waste in our society. In order to clearly describe the multifunction products, they are divided into four categories in this chapter. The first kind of multifunction product is that multiple functions can be performed at the same time; the second kind is that multiple functions can be performed by changing different working status; the third kind is that multiple functions can be performed by reconfiguring components; the fourth kind is that multiple functions can be realized through changing different executive components on the basis of a common component.

Many methods are well equipped for use with some design problems. But few computational tools are specialized for the multifunction product. Therefore, it is more and more important to specially investigate the design approach of the multifunction product [1, 2]. Matrix is a stylized tool and widely used in product conceptual design. It is a good tool mapping information flow and its impact in product design processes, and it can represent visually the network of interactions among design functions and facilitates analysis of these interactions. Based on the product matrix, the objective of this study is to develop a matrix-based stylized bottom-up approach aiming at generating a conceptual solution of all categories of multifunction product. In the approach, a few matrices are proposed to analyze the reconfiguration of two single-function products.

The chapter is outlined as follows. Section 3.2 introduces the related researches of product design, followed by the construction of function models in Sect. 3.3. The concept generation approach of the multifunction product is introduced in Sect. 3.4. Finally, application of the proposed method for a shaver design is demonstrated.

D. Tang et al., *Matrix-Based Product Design and Change Management*,
DOI 10.1007/978-981-10-5077-0_3

3.2 Related Research of Multifunction Product Design

Multifunction product is defined as complex function product by Feng et al. [3]. They have proposed a conceptual design cycle mapping model and studied the cycle solving process and realization for conceptual design of the multifunction product. Hu et al. [4] have incorporated the fuzzy set theory into the reconfigurable design and proposed a fuzzy reconfigurable design method of multifunction products. While these two approaches are useful in the top-down design of new multifunction products, they do not address the bottom-up combination of the existing products. Li has given a design method of the multifunction product family, and an integrated product family reconfigurable design system is developed and applied in the design process of air separation system [5]. But the presented method is effective only for the multifunction product family. Based on the similarity theory, Zhang et al. have constructed a design platform of multifunction products [6]. Liu et al. [7] have proposed a bottom-up platform design method of the multifunction product. Wei et al. [8] have presented a modular design method of multifunction product based on the product module. However, these three approaches may be effective only for the similar multifunction products.

Recently, a new design principle named as Transformational Design Theory has been proposed in [1]. It provides an avenue for developing transforming systems. In [9], the transformation design is refined to be simpler, easier to learn and follow, and more effective in drawing out untapped potential from the design process. In addition, a research presents a twofold process known as the Transformer Diagram Matching Method and shows the results on a fully functioning prototype of an office supply transformer [2]. The process takes transformation design methodology a large step forward by bridging its two ways of design together, directive and intuitive. According to the categories of the multifunction product, the transforming systems or products shall be the third kind of multifunction product, and the principle is appropriate for this kind of multifunction product. But for other categories of multifunction product, it is not known whether the transformational design theory is effective or not. Furthermore, mathematical design tools for other categories of multifunction product are still needed.

Matrix is a stylized tool and widely used in product conceptual design. By applying different matrix operations, the design activities can be organized and analyzed [10]. One example of a matrix-based method is the House of Quality from Quality Function Deployment (QFD), where customer requirements are mapped to engineering characteristics [11]. Other matrix-based methods include the incidence matrix [12], the design structure matrix [13], and the function impact matrix [14]. These matrixes are best suited for specific domains or applications of the product conceptual design. But there is a lack of matrix-based methods to support the combination of two or more existing single-function products or subsystems into a new multifunction product or system. Thus this study is carried out mainly to bridge the gap in matrix-based conceptual solution generation of multifunction product.

3.3 Function Model

The concept design is a complicated thinking activity and the function model is the basis of the concept design [15]. The function model also increases the clarity of the design problem by tracking of the input and output flows. It is used to capture the design knowledge from the existing products and represent a form-independent blueprint of a product.

3.3.1 *Functional Basis*

In order to describe the product function, Stone and Wood proposed a language named as functional basis [16]. The functional basis, where product function is characterized in a verb-object (function-flow) format, is a consistent language or coding system required to ensure that others can read it. It is intended to comprehensively describe the mechanical design space without repetition. The functional basis contains three primary flows and eight primary functions. The three primary flows are material, signal, and energy flow. The material has five further specified secondary categories with an expanded list of tertiary categories. The signal has two further specified secondary categories with an expanded list of tertiary categories. The energy has thirteen further specified secondary categories with an expanded list of tertiary categories, as shown in Table 3.1. Eight primary functions are "branch", "channel", "connect", "control magnitude", "convert", "provision", "signal" and "support". Each primary function has several further specified secondary categories with an expanded list of tertiary categories, as shown in Table 3.2. The clear definitions have been developed for all flows and functions in [17]. The functional basis is applied to the areas of product architecture development, functional model generation, and design information transmittal [18].

3.3.2 *Construction of Function Model*

The product function is a description of the design system. In order to construct the function model and analyze the relationships between energy, material, or signal of

Table 3.1 Example of the expanded flow

Primary class	Secondary	Tertiary	Correspondents
Energy	Electrical	Electrical	Electromotive force, current
	Mechanical	Rotational energy	Torque, angular velocity
		Translational energy	Force, linear velocity
	Pneumatic	Pneumatic	Pressure, mass flow

Table 3.2 Example of the expanded function

Primary class	Secondary	Tertiary	Correspondents
Channel	Import	Import	Form entrance, allow, input, and capture
	Export	Export	Dispose, eject, emit, empty
			Remove, destroy, and eliminate
	Transfer	Transport	Advance, lift, and move
		Transmit	Conduct, convey

the product conveniently, the overall function of a product can be decomposed into a number of sub-functions. The function of a product can be described by the sub-function sets. The sub-functions are the roles' abstractions of the existing parts or process. They are described by the functional basis and are used to construct the function model. The function model of any products can be generated by this approach. The steps of the functional model construction are as follows:

1. The overall function and the input/output flow of the product are confirmed.
2. The overall function is decomposed into sub-functions described by the functional basis.
3. The functional chains of each input flow are constructed.
4. The function model of a product is acquired by interlinking all functional chains [7, 19].

3.4 Conceptual Solution Generation Approach (CSGA) of Multifunction Product

The first mission of the proposed approach is the function analysis and the function model construction of two single-function products. Second, the sub-function similarities of two products are analyzed and the functional similarity matrix (FSM) is constructed. Third, the component–component matrix (CCM) of two single-function products is acquired by calculating the FSM and the function–component matrix (FCM). Finally, the conceptual solution of the new multifunction product is generated by analyzing the components' relationship in the CCM. In brief, the flow chart of the approach is shown in Fig. 3.1.

3.4.1 Function Analysis

The function analysis is the first mission in the proposed approach. It includes construction of the functional model, identification of the chief sub-functional

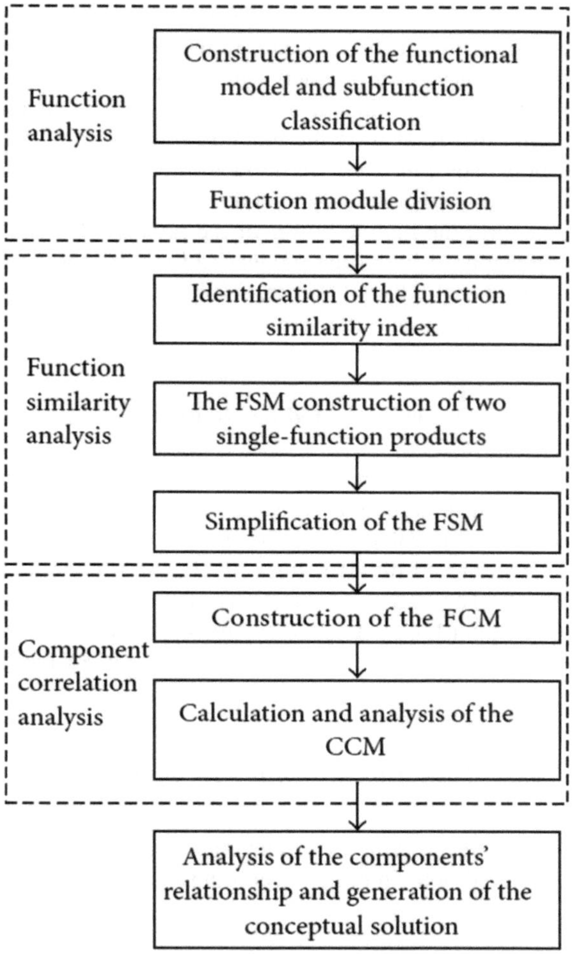

Fig. 3.1 Flowchart of the proposed method

chain, sub-function classification, and function module division. The construction of the functional model has been introduced in Sect. 3.3, and others will be introduced in the following subsections.

3.4.1.1 Chief Sub-functional Chain Identification and Sub-function Classification

The sub-functional chain is a continuous sub-function set of a product and it may be the energy, material, or signal flow. The sub-functional chains are generated from the functional model by the sequence of the flow. If there are multiple sub-functional chains, a chief functional chain shall be chosen and analyzed primarily. Generally, the chief sub-functional chain is the sub-functional chain that contains most sub-functions. If the chief functional chains of two products are similar, these two products can be integrated into a new product. Otherwise, the integration feasibility of these two products is smaller.

In order to analyze the function similarities of two single function products, the product sub-functions are divided into three categories: basic function, application function, and accessory function. The basic functions are used to transfer and transform the motive power of the product. The application function indicates the application value of the product and is a function set used to distinguish a product with other products. The accessory function is a discrete function set besides the basic function and the application function.

3.4.1.2 Functional Module Division

In order to compare and analyze the similarity between the sub-functions of two products, the function model is divided into different modules on the basis of the sub-function categories. The basic functions and the application functions are sequential and divided, respectively, into different modules in the sub-functional chains. The accessory functions are not coterminous like the basic functions or application functions, so they are collected into a single module. The sub-function similarity is analyzed only in the same type of the functional module.

3.4.2 Functional Similarity Analysis

In this subsection, the functional similarity index is defined and the FSM is structured and simplified according to the proposed principles.

3.4.2.1 Functional Similarity Index

The product similarity is introduced in [20]. In their opinion, if there are one or more same important functions in different products, these products are function similar. In [8], the product similarity is analyzed from two aspects: property and characteristic, relation and function. In order to quantify the function similarity, the similarity index is given to represent the similarity and it is classed into seven degrees: 0, 0.3, 0.5, 0.7, 0.9, 1, and −1. The similarity index of a sub-function pair is confirmed by identifying the function/flow descriptors about the primary class, the secondary class, and the correspondents, as shown in Tables 3.1 and 3.2. As we move from the primary class to the correspondents, the level of abstraction decreases and the functions become more and more specific in nature. The higher the level of similarity between two sub-functions at all three classes of function and flow sets, the bigger the similarity index of the two sub-functions will be.

Function similarity matrix			Product B												
			Basic functions					Application functions				Accessory functions			
			F_1^b	F_2^b	F_3^b	F_4^b	·	F_6^b	F_7^b	F_8^b	·	F_{10}^b	F_{11}^b	·	·
Product A	Basic functions	F_1^a	0.9	1	0	0	0								
		F_2^a	0	0	0.7	0	0								
		F_3^a	0	0.3	0	0	1								
		·	·	·	·	·	·								
		·	·	·	·	·	·								
	Application functions	F_6						1	0	0	0				
		F_7						0	0.5	0	0				
		·						·	·	·	·				
		·						·	·	·	·				
	Accessory functions	F_{10}^a										0.7	0	0	0
		F_{11}^a										0	−1	0	0
		F_{12}^a										0	0	1	0
		·										·	·	·	·
		·										·	·	·	·

Fig. 3.2 Function similarity matrix

3.4.2.2 Functional Similarity Matrix (FSM)

In order to analyze the functional similarity of two existing products, the FSM is constructed, as shown in Fig. 3.2. In the matrix, the sub-functions are ranked by the sequence of the flow in the functional model. The function similarity is analyzed only between same types of the modules. The entry in each square is the similarity index of a sub-function pair.

The entry in the matrix, m_{ij}, indicates the similarity level of a sub-function pair. These entries can assume the following values:

1. $m_{ij} = 1$ indicates that the corresponding sub-functions i and j are the same.
2. $m_{ij} = -1$ indicates that the corresponding sub-functions i and j are conflicting.
3. $m_{ij} = 0$ indicates that there is no similarity between the functions i and j.
4. $m_{ij} \in [0.3, 0.5, 0.7, 0.9]$ indicates that there is partial similarity between the functions i and j.

3.4.2.3 Simplification of the FSM

In order to avoid a high degree of internal function coupling in the new product, it is assumed that the similarity relationship only exists between two sub-functions and the corresponding sub-function pair complies with the positive sequence of the flow. The entries that do not conform to the assumption shall be removed by setting

Fig. 3.3 An original function similarity matrix

FSM		Product B				
		F_1^b	F_2^b	F_3^b	F_4^b	F_5^b
Product A	F_1^a	−1	0	0	0	0
	F_2^a	0	0	0.3	0	0
	F_3^a	0	0.7	0	0	0
	F_4^a	0	0	0	0	0.3
	F_5^a	0	0	0	0	1

Fig. 3.4 A simplified function similarity matrix

FSM		Product B				
		F_1^b	F_2^b	F_3^b	F_4^b	F_5^b
Product A	F_1^a	−1	0	0	0	0
	F_2^a	0	0	0	0	0
	F_3^a	0	0.7	0	0	0
	F_4^a	0	0	0	0	0
	F_5^a	0	0	0	0	1

to 0. This leads us to the following principle: the values on the diagonal and the higher values below the diagonal shall be preferentially reserved, and this keeps the nonzero values in the FSM to the right and downward trends. Therefore, any similarity relationship between more than two sub-functions must be parsed further, and the feedback and the dual relationship are eliminated by setting the lowest entry to zero. For instance, as shown in Fig. 3.3, F_5^b has the similarity relationships F_4^a with F_4^a and F_5^a, and the presence of similarity relationships between F_2^a and F_3^b and F_3^a and F_2^b indicate a reversal of flow since it would require a feedback mechanism. Therefore, the value 0.3 corresponding to F_2^a and F_3^b and the value 0.3 corresponding to F_4^a and F_5^b shall be set to zero according to the principle, as shown in Fig. 3.4. For easy calculation in the next section, the matrix in Fig. 3.4 is transformed into Eq. (3.1).

$$M_{\mathrm{FSM}} = \begin{array}{c} \\ F_1^a \\ F_2^a \\ F_3^a \\ F_4^a \\ F_5^a \end{array} \begin{array}{c} \begin{array}{ccccc} F_1^b & F_2^b & F_3^b & F_4^b & F_5^b \end{array} \\ \begin{pmatrix} -1 & 0 & 0 & 0 & 0 \\ 0 & 0 & 0 & 0 & 0 \\ 0 & 0.7 & 0 & 0 & 0 \\ 0 & 0 & 0 & 0 & 0 \\ 0 & 0 & 0 & 0 & 1 \end{pmatrix} \end{array} \tag{3.1}$$

3.4.3 Component Correlation Analysis

In this subsection, the FCM of every single-function product is structured. The CCM can be acquired by multiplying FCM and FSM, and then the component relationships in CCM can be analyzed.

3.4.3.1 Function Component Matrix (FCM)

In order to convert the function similarity relationships in FSM to the components correlations, the FCM of each product is structured. As shown in Fig. 3.5, two hypothetical FCMs of products A and B are constructed. The columns of the FCM^a must be in the same order as the columns of the FSM as shown in Fig. 3.4, and the columns of the FCM^b must be in the same order as the rows of the FSM. But the arrangement of components in the FCM does not affect the final results of this phase. This is because the propagation of the function similarity from the FSM to the components of the FCM occurs only through the sub-functions of the products.

3.4.3.2 Calculation and Analysis of the Component–Component Matrix

In order to use a matrix to describe the components' correlation of two products, the CCM is presented and acquired by Eq. (3.2).

$$[\mathrm{FCM}^a]_{mi}^T \times [\mathrm{FSM}]_{ij} \times [\mathrm{FCM}^b]_{jn} = [\mathrm{CCM}]_{mn} \tag{3.2}$$

where m and n indicate the total components' number of product A and product B, respectively. This equation is used to map the similarity of the function pair in the

Product A					
FCM^a	C_1^a	C_2^a	C_3^a	C_4^a	C_5^a
F_1^a	1	0	0	0	0
F_2^a	0	1	0	0	0
F_3^a	0	0	1	0	0
F_4^a	0	0	0	1	0
F_5^a	0	0	0	1	1

Product B					
FCM^b	C_1^b	C_2^b	C_3^b	C_4^b	C_5^b
F_1^b	1	0	0	0	0
F_2^b	0	1	0	0	0
F_3^b	0	0	1	0	0
F_4^b	0	0	0	1	0
F_5^b	0	0	0	1	1

Fig. 3.5 Function–component matrix

FSM into the CCM. The calculation process of the FSM in Fig. 3.4 mapped into a CCM is shown as follows:

$$
\begin{array}{c}
 \\ C_1^a \\ C_2^a \\ C_3^a \\ C_4^a \\ C_5^a
\end{array}
\begin{array}{c}
\begin{array}{ccccc} F_1^a & F_2^a & F_3^a & F_4^a & F_5^a \end{array} \\
\begin{pmatrix} 1 & 0 & 0 & 0 & 0 \\ 0 & 1 & 0 & 0 & 0 \\ 0 & 0 & 1 & 0 & 0 \\ 0 & 0 & 0 & 1 & 1 \\ 0 & 0 & 0 & 0 & 1 \end{pmatrix}
\end{array}
*
\begin{array}{c}
 \\ F_1^a \\ F_2^a \\ F_3^a \\ F_4^a \\ F_5^a
\end{array}
\begin{array}{c}
\begin{array}{ccccc} F_1^b & F_2^b & F_3^b & F_4^b & F_5^b \end{array} \\
\begin{pmatrix} -1 & 0 & 0 & 0 & 0 \\ 0 & 1 & 0 & 0 & 0 \\ 0 & 0.7 & 1 & 0 & 0 \\ 0 & 0 & 0 & 0 & 0 \\ 0 & 0 & 0 & 0 & 1 \end{pmatrix}
\end{array}
*
\begin{array}{c}
 \\ F_1^b \\ F_2^b \\ F_3^b \\ F_4^b \\ F_5^b
\end{array}
\begin{array}{c}
\begin{array}{ccccc} C_1^b & C_2^b & C_3^b & C_4^b & C_5^b \end{array} \\
\begin{pmatrix} 1 & 0 & 0 & 0 & 0 \\ 0 & 1 & 0 & 0 & 0 \\ 0 & 0 & 1 & 0 & 0 \\ 0 & 0 & 0 & 1 & 0 \\ 0 & 0 & 0 & 1 & 1 \end{pmatrix}
\end{array}
$$

$$
=
\begin{array}{c}
 \\ C_1^a \\ C_2^a \\ C_3^a \\ C_4^a \\ C_5^a
\end{array}
\begin{array}{c}
\begin{array}{ccccc} F_1^b & F_2^b & F_3^b & F_4^b & F_5^b \end{array} \\
\begin{pmatrix} -1 & 0 & 0 & 0 & 0 \\ 0 & 0 & 0 & 0 & 0 \\ 0 & 0.7 & 0 & 0 & 0 \\ 0 & 0 & 0 & 0 & 0 \\ 0 & 0 & 0 & 0 & 1 \end{pmatrix}
\end{array}
*
\begin{array}{c}
 \\ F_1^b \\ F_2^b \\ F_3^b \\ F_4^b \\ F_5^b
\end{array}
\begin{array}{c}
\begin{array}{ccccc} C_1^b & C_2^b & C_3^b & C_4^b & C_5^b \end{array} \\
\begin{pmatrix} 1 & 0 & 0 & 0 & 0 \\ 0 & 1 & 0 & 0 & 0 \\ 0 & 0 & 1 & 0 & 0 \\ 0 & 0 & 0 & 1 & 0 \\ 0 & 0 & 0 & 1 & 1 \end{pmatrix}
\end{array}
=
\begin{array}{c}
 \\ C_1^a \\ C_2^a \\ C_3^a \\ C_4^a \\ C_5^a
\end{array}
\begin{array}{c}
\begin{array}{ccccc} C_1^b & C_2^b & C_3^b & C_4^b & C_5^b \end{array} \\
\begin{pmatrix} -1 & 0 & 0 & 0 & 0 \\ 0 & 0 & 0 & 0 & 0 \\ 0 & 0.7 & 0 & 0 & 0 \\ 0 & 0 & 0 & 0 & 0 \\ 0 & 0 & 0 & 0 & 1 \end{pmatrix}
\end{array}
\tag{3.3}
$$

The operator * in (3.2) and (3.3) has two calculation steps as follows:

Step1: When the row i of the first matrix (e.g., FCM^a) is multiplied by the column j of the second matrix (e.g., FSM), each element of the row i is multiplied by the corresponding element of the column j, and the results are ranked in a set $\{T_{ij}| m_{i1} \times m_{1j}, m_{i2} \times m_{2j}, \ldots, m_{ix} \times m_{xj}\}$, where m_{ix} represents the element of the row i in the first matrix and m_{xj} represents the element of the column j in the second matrix. For instance, the calculation process of the row C_1^a in the first matrix multiplied by the column F_1^b in the second matrix Eq. (3.3) is shown as follows:

$$
\begin{aligned}
C_1^a \times F_1^b &= [1,0,0,0,0] \times [-1,0,0,0,0] \\
&= \{1 \times -1, 0 \times 0, 0 \times 0, 0 \times 0, 0 \times 0\} \\
&= \{-1,0,0,0,0\}
\end{aligned}
\tag{3.4}
$$

Step2: The element m_{ij} of the new matrix is selected from the results of *Step 1* according to the following rules

$$
M_{ij} = \begin{cases} 0 & T_{ij} = \{0\}, \\ -1 & if - 1 \in T_{ij}, \\ z & if - 1 \notin T_{ij}, z\,\text{is the minimum in}\, T_{ij}. \end{cases}
\tag{3.5}
$$

where the first condition identifies no relationship between two elements. The second condition identifies the conflicting relationship. The third condition ensures that only the minimum in T_{ij} is taken into consideration when there are multiple positive numbers. The lowest number conservatively reflects the minimum similarity.

Fig. 3.6 Component–component matrix of two products

CCM		Product B				
		C_1^b	C_2^b	C_3^b	C_4^b	C_5^b
Product A	C_1^a	−1	0	0	0	0
	C_2^a	0	0	0	0	0
	C_3^a	0	0.7	0	0	0
	C_4^a	0	0	0	1	1
	C_5^a	0	0	0	1	1

According to the above principles, the result of C_1^a multiplied by F_1^b is -1, as shown by the following Eq. (3.6). Consequently, the results of (3.3) are acquired and they are presented visually in CCM as shown in Fig. 3.6.

$$\begin{aligned} C_1^a \times F_1^b &= [1,0,0,0,0] \times [-1,0,0,0,0] \\ &= \{-1,0,0,0,0\} \\ &= -1 \end{aligned} \tag{3.6}$$

The components' relationships of two products are indicated in CCM. Different types of the values in CCM are as follows. The value -1 indicates that the components C_1^a and C_1^b are conflicting. The value 0 identifies that the components C_2^a and C_2^b are dormant and they are not used in the new product. The values between 0 and 1 indicate that the components have the feasibility to be redesigned into a single component that can perform two functions of the existing products. The value 1 identifies that the components C_4^a, C_4^b, C_5^a, and C_5^b have the same function and can be replaced by each other.

By analyzing the components' relationships in CCM, a few components of two existing products will be reserved in the new product, and a few components will be modified for the new application, and the others will be eliminated. Consequently, the conceptual solution of the new product can be acquired. The detailed application procedure is illustrated in the following section.

3.5 Application of CSGA for Multifunction Product

3.5.1 *Function Analysis*

Two single-function products, a shaver and a portable cleaner, are shown in Fig. 3.7. The overall functions of the two products are, respectively, decomposed into a number of sub-functions that are described by the functional basis, and the function models of the two single-function products are constructed, respectively, as shown in Figs. 3.8 and 3.9.

(b) Shaver (a) Portable cleaner

Fig. 3.7 Single-function products

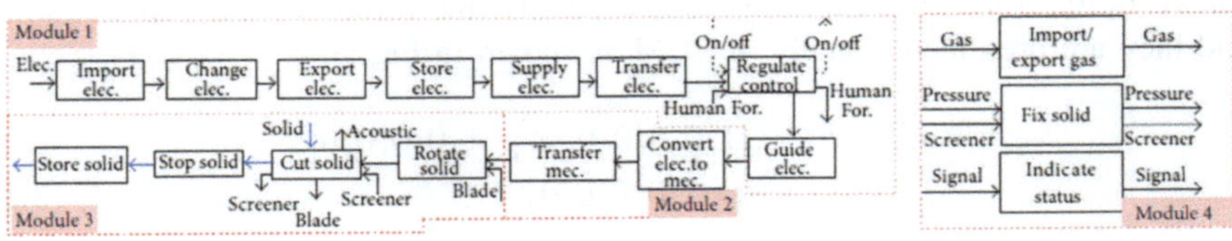

(a) Chief function chain (b) Accessory functions

Fig. 3.8 Function model of the shaver

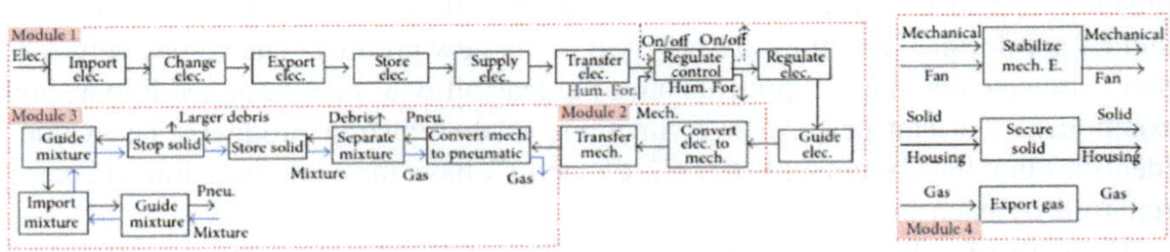

(a) Chief function chain (b) Accessory functions

Fig. 3.9 Function model of the portable cleaner

In Figs. 3.8 and 3.9, the function models of the two products are, respectively, divided into four modules according to the types of the energy, material, or signal flow. For instance, in Fig. 3.8, the flow in module 1 is the electrical energy flow. The flow of module 2 is the mechanical energy flow, and the flow in module 3 is the solid material flow. The accessory functions are regarded as an independent module. These modules are divided into three categories according to the sub-function classification introduced in Sect. 3.4.1.1. The categories of the modules are listed in Table 3.3.

As shown in Table 3.3, module 1 and module 2 of two products belong to the same functional category. Module 3 is a material flow module including a solid material flow and a mixture material flow. This indicates that the application function is the key distinction of two products. Therefore, the sub-functions in the module 3 are primarily modified for the function of the new product.

Table 3.3 Categories of the modules

Module category	Product 1: shaver	Product 2: portable cleaner
Basic function	Electrical energy module (1)	Electrical energy module (1)
Basic function	Mechanical energy module (2)	Mechanical energy module (2)
Application function	Material flow module (3)	Material flow module (3)
Accessory function	Accessory function module (4)	Accessory function module (4)

3.5.2 *Functional Similarity Analysis*

The FSMs between the shaver and the portable cleaner are structured according to their function models, as shown in Fig. 3.10. If two sub-functions are conflicting, the corresponding entry is −1, and it does not exist in Fig. 3.10. Other similarity indexes of the function pairs are represented from 0 to 1 according to the similarity degree. The FSM is simplified according to the principle introduced in Sect. 3.4.2.3, and the result is shown in Fig. 3.11. Almost all many-to-many relationships of the sub-function pairs are eliminated, and one-to-one relationships are reserved in the simplified FSM. Only the shaver's sub-function *Guide electrical energy* corresponds to two sub-functions of the portable cleaner: *Regulate electrical energy* and *Guide electrical energy*. In addition, some sub-functions correspond to nothing, such as *Stop solid*, *Indicate status*, and *Guide mixture*.

Function similarity matrix (FSM)				Portable cleaner																				
				Basic functions											Application functions							Accessory functions		
				Import elec.	Change elec.	Export elec.	Store elec.	Supply elec.	Transfer elec.	Regulate con.	Regulate elec.	Guide elec.	Convert elec. to mech.	Transfer mech.	Convert mech. to pneu.	Separate mix.	Store solid	Stop solid	Guide mixture	Import mixture	Guide mixture	Stabilize mech.	Secure solid	Export gas
				1	2	3	4	5	6	7	8	9	10	11	12	13	14	15	16	17	18	19	20	21
Shaver	Basic functions	Import elec.	1	1	0	0	0	0	0	0	0	0	0	0										
		Change elec.	2	0	1	0	0	0	0	0	0.7	0	0	0										
		Export elec.	3	0	0	1	0	0	0.7	0	0	0.7	0	0										
		Store elec.	4	0	0	0	1	0	0	0	0	0	0	0										
		Supply elec.	5	0	0	0	0	1	0.7	0	0	0.7	0	0										
		Transfer elec.	6	0	0	0.7	0	0.7	1	0	0	1	0	0										
		Regulate control	7	0	0	0	0	0	0	1	0	0	0	0										
		Guide elec.	8	0	0	0.7	0	0.7	1	0	0	1	0	0										
		Convert elec. to mech.	9	0	0	0	0	0	0	0	0	0	1	0										
		Transfer mech.	10	0	0	0	0	0	0	0	0	0	0	1										
	Appli. functions	Rotate solid	11												1	0	0	0	0	0	0			
		Cut solid	12												0	0.7	0	0.3	0	0	0			
		Stop solid	13												0	0.7	0.3	1	0.7	0	0.7			
		Store solid	14												0	0	1	0.3	0.7	0	0.7			
	Acce. fun.	Fix solid	15																			0	1	0
		Indicate status	16																			0	0	0
		Import/exp. gas	17																			0	0	1

Fig. 3.10 FSM between the shaver and the portable cleaner

Filtered total function similarity matrix (FTFSM)				Portable cleaner																				
				Basic functions											Application functions							Accessory functions		
				Import elec.	Change elec.	Export elec.	Store elec.	Supply elec.	Transfer elec.	Regulate con.	Regulate elec.	Guide elec.	Convert elec. to mech.	Transfer mech.	Convert mech. to pneu.	Separate mix.	Store solid	Stop solid	Guide mixture	Import mixture	Guide mixture	Stabilize mech.	Secure solid	Export gas
				1	2	3	4	5	6	7	8	9	10	11	12	13	14	15	16	17	18	19	20	21
Shaver	Basic functions	Import elec.	1	1	0	0	0	0	0	0	0	0	0	0										
		Change elec.	2	0	1	0	0	0	0	0	0	0	0	0										
		Export elec.	3	0	0	1	0	0	0	0	0	0	0	0										
		Store elec.	4	0	0	0	1	0	0	0	0	0	0	0										
		Supply elec.	5	0	0	0	0	1	0	0	0	0	0	0										
		Transfer elec.	6	0	0	0	0	0	1	0	0	0	0	0										
		Regulate control	7	0	0	0	0	0	0	1	0	0	0	0										
		Guide elec.	8	0	0	0	0	0	0	0	0.7	1	0	0										
		Convert elec. to mech.	9	0	0	0	0	0	0	0	0	0	1	0										
		Transfer mech.	10	0	0	0	0	0	0	0	0	0	0	1										
	Appli. functions	Rotate solid	11												0.9	0	0	0	0	0	0			
		Cut solid	12												0	0.7	0	0	0	0	0			
		Stop solid	13												0	0	0	0	0	0	0			
		Store solid	14												0	0	1	0	0	0	0			
	Acce. fun.	Fix solid	15																			0	1	0
		Indicate status	16																			0	0	0
		Import/exp. gas	17																			0	0	1

Fig. 3.11 Simplified FSM between the shaver and the portable cleaner

3.5.3 Component Correlation Analysis

The FCMs of the shaver and the portable cleaner are, respectively, constructed according to the relationships between the sub-functions and components, as shown in Figs. 3.12 and 3.13, respectively. The relationship between the sub-function and component is represented by 0 or 1, and it is not further distinguished. The entry 1 indicates that the sub-function can be performed by the component on the column in the FCM. The entry 0 indicates that the sub-function is irrelevant with the component on the column of FCM. The CCM of the shaver and the portable cleaner is calculated by the FSM multiplied by FCMs of the shaver and the cleaner according to Eq. (3.3), and the result is shown in Fig. 3.14.

3.6 Results and Discussion

The entry 1 in the CCM shown in Fig. 3.14 indicates the maximal correlation degree of the component pair. The entry 0 indicates that there is no correlation for the components in the row and the column. The entries between 0 and 1 indicate the correlation degree of two components. Therefore, the correlation is divided into three kinds: no-correlation, simple-correlation, and complex-correlation. The correlation classifications are shown in Table 3.4, and the complex correlations are shown in Fig. 3.15.

FCM of shaver		Charger	Battery	Wire	Switch	Diode	Motor	Motor shaft	Blade bearing	Blade	Screener	Housing	Blade cabin	Luminous diode
		1	2	3	4	5	6	7	8	9	10	11	12	13
Import elec.	1	1	0	0	0	0	0	0	0	0	0	0	0	0
Change elec.	2	1	0	0	0	0	0	0	0	0	0	0	0	0
Export elec.	3	1	0	0	0	0	0	0	0	0	0	0	0	0
Supply elec.	4	0	1	0	0	0	0	0	0	0	0	0	0	0
Store elec.	5	0	1	0	0	0	0	0	0	0	0	0	0	0
Transfer elec.	6	0	0	1	0	0	0	0	0	0	0	0	0	0
Regulate cont.	7	0	0	0	1	0	0	0	0	0	0	0	0	0
Transfer elec.	8	0	0	0	0	1	0	0	0	0	0	0	0	0
Convert elec. to mech.	9	0	0	0	0	0	1	0	0	0	0	0	0	0
Transfer mech.	10	0	0	0	0	0	0	1	0	0	0	0	0	0
Rotate solid	11	0	0	0	0	0	0	0	1	0	0	0	0	0
Cut solid	12	0	0	0	0	0	0	0	0	1	1	0	0	0
Stop solid	13	0	0	0	0	0	0	0	0	0	0	1	0	0
Store solid	14	0	0	0	0	0	0	0	0	0	0	0	1	0
Fix solid	15	0	0	0	0	0	0	0	0	0	1	0	0	0
Indicate status	16	0	0	0	0	0	0	0	0	0	0	0	0	1
Imp./exp. gas	17	0	0	0	0	0	0	0	0	0	1	0	1	0

Fig. 3.12 FCM of the shaver

The no-correlation components are eliminated in the conceptual solution generation of the new product, and these components are 15% of the total components. If the correlation index of a simple-correlation component pair is 1, the component pair is irrelevant with other component pairs, and a component in the pair can be replaced by another one. These components do not need to be modified and are the core components of the new product. These components account for 37% of the total components. If the correlation index of the simple-correlation component pair is not 1, this indicates that the two components in the pair are similar. The two

FCM of portable cleaner		Adapter	Battery	Contact plate and wire	Switch	Diode	Motor	Motor shaft	Impeller	Filter bag	Rubber flap	Rubber flap holder	Nozzle	Exhaust vents	Housing
		1	2	3	4	5	6	7	8	9	10	11	12	13	14
Import elec.	1	1	0	0	0	0	0	0	0	0	0	0	0	0	0
Change elec.	2	1	0	0	0	0	0	0	0	0	0	0	0	0	0
Export elec.	3	1	0	0	0	0	0	0	0	0	0	0	0	0	0
Supply elec.	4	0	1	0	0	0	0	0	0	0	0	0	0	0	0
Store elec.	5	0	1	0	0	0	0	0	0	0	0	0	0	0	0
Transfer elec.	6	0	0	1	0	0	0	0	0	0	0	0	0	0	0
Regulate cont.	7	0	0	0	1	0	0	0	0	0	0	0	0	0	0
Regulate elec.	8	0	0	0	1	0	0	0	0	0	0	0	0	0	0
Guide elec.	9	0	0	0	0	1	0	0	0	0	0	0	0	0	0
Convert elec. to mech.	10	0	0	0	0	0	1	0	0	0	0	0	0	0	0
Transfer mech.	11	0	0	0	0	0	0	1	0	0	0	0	0	0	0
Convert mech. to pneu.	12	0	0	0	0	0	0	0	1	0	0	0	0	0	0
Separate mixture	13	0	0	0	0	0	0	0	0	1	0	0	0	0	0
Store solid	14	0	0	0	0	0	0	0	0	1	0	0	0	0	0
Stop solid	15	0	0	0	0	0	0	0	0	0	1	0	0	0	0
Guide mixture	16	0	0	0	0	0	0	0	0	0	0	1	0	0	0
Import mixture	17	0	0	0	0	0	0	0	0	0	0	0	1	0	0
Guide mixture	18	0	0	0	0	0	0	0	0	0	0	0	1	0	0
Stabilize mech.	19	0	0	0	0	0	0	0	0	0	0	0	0	0	1
Secure solid	20	0	0	0	0	0	0	0	0	0	0	0	0	0	1
Export gas.	21	0	0	0	0	0	0	0	0	0	0	0	0	1	1

Fig. 3.13 FCM of the portable cleaner

components can be modified and integrated into a component of the new product that performs two similar sub-functions of the existing single-function products. These components account for 7% of the total components. For the complex-correlation components, if one of them is modified, the others will be affected, as shown in Fig. 3.15. Consequently, these components will be considered as a whole and coherently redesigned one with others in the new conceptual solution. These components account for 41% of the total components.

CCM			Portable cleaner													
			Adapter	Battery	Contact plate and wire	Switch	Diode	Motor	Motor shaft	Impeller	Filter bag	Rubber flap	Rubber flap holder	Nozzle	Exhaust vents	Housing
			1	2	3	4	5	6	7	8	9	10	11	12	13	14
Shaver	Charger	1	1	0	0	0	0	0	0	0	0	0	0	0	0	0
	Battery	2	0	1	0	0	0	0	0	0	0	0	0	0	0	0
	Wire	3	0	0	1	0	0	0	0	0	0	0	0	0	0	0
	Switch	4	0	0	0	1	0	0	0	0	0	0	0	0	0	0
	Diode	5	0	0	0	0.7	1	0	0	0	0	0	0	0	0	0
	Motor	6	0	0	0	0	0	1	0	0	0	0	0	0	0	0
	Motor shaft	7	0	0	0	0	0	0	1	0	0	0	0	0	0	0
	Blade bearing	8	0	0	0	0	0	0	0	0.9	0	0	0	0	0	0
	Blade	9	0	0	0	0	0	0	0	0	0.7	0	0	0	0	0
	Screener	10	0	0	0	0	0	0	0	0	0.7	0	0	0	1	1
	Housing	11	0	0	0	0	0	0	0	0	0	0	0	0	0	1
	Blade cabin	12	0	0	0	0	0	0	0	0	1	0	0	0	1	1
	Luminous diode	13	0	0	0	0	0	0	0	0	0	0	0	0	0	0

Fig. 3.14 CCM of the shaver and the portable cleaner

The existing single-function product that includes fewer no-correlation components is used for basic architecture of the new product, and this can reduce the ambiguity of the new architecture caused by removing more components. The shaver has a no-correlation component and the portable cleaner has three no-correlation components; therefore, the shaver is used for the basic architecture of the new product. Consequently, the no-correlation component can be removed, such as nozzle. The blade bearing and the impeller are integrated into a component of the new product that performs two sub-functions, *Rotate solid* and *Convert mechanical energy to pneumatic energy*, at the same time. A few components are reserved and used in the new product, and they are motor, battery, adapter, wire, and motor shaft. The blade cabin and the housing are redesigned into a new blade cabin and a dust cabin according to the complex-correlation relationship. On the basis of the above, the conceptual solution of the new product is acquired, as shown in Fig. 3.16. The new product has two functions at the same time: shaving and residue absorption. The beard residue can be collected into an easy dismantling cabin and cleared easily. This cabin is named residue cabin, as shown in Fig. 3.16.

Table 3.4 Categories of the modules

	Shaver	Portable cleaner	Correlation index
No-correlation component	Luminous diode (13)	Rubber flap (10)	0
		Rubber flap holder (11)	
		Nozzle (12)	
Simple-correlation components	Charger (1)	Adapter (1)	1
	Battery (2)	Battery (2)	1
	Wire (3)	Contact plate and wire (3)	1
	Motor (6)	Motor (6)	1
	Motor shaft (7)	Motor shaft (7)	1
	Blade bearing (8)	Impeller (8)	0.9
Complex-correlation components	Switch (4)	Switch (4)	
	Diode (5)	Diode (5)	
	Blade (9)		
	Screener (10)	Filter bag (9)	
	Housing (11)	Exhaust vents (13)	
	Blade cabin (12)	Housing (14)	

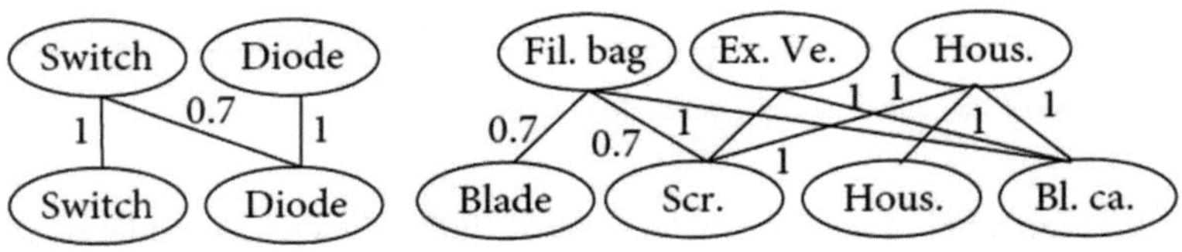

(a) First set (b) Second set

Fig. 3.15 Complex correlations

3.7 Conclusion

In this chapter, a conceptual solution generation approach is proposed. It is a concise, stylized, and bottom-up approach and specialized for the multifunction product. In this approach, the overall function of the existing single-function product is decomposed into a number of sub-functions described by the functional basis, and this allows the functions of the different products to be compared and analyzed at the same level. A few matrixes are used to analyze the sub-functions' similarity and the components' correlation of two products, and the stylization, quantification, and automation of the calculation process are easily realized. Using the approach, two single-function products can be combined into a multifunction product. By similar way, a single-function product can be combined with a sub-function chain of a multifunction product, or two sub-function chains of

Fig. 3.16 Conceptual solution of the new product

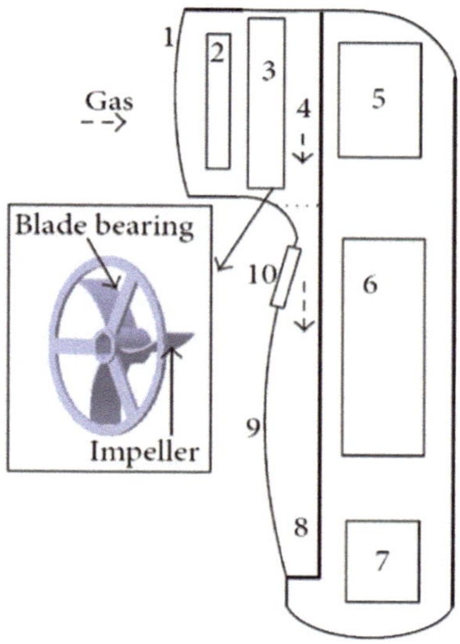

(6) Screener
(7) Blade
(8) Fit of the blade base and impeller
(9) Blade cabin
(10) Motor

(1) Battery
(2) Adapter
(3) Residue cabin
(4) Quick disassembling baffle
(5) Exhaust vent

different products can be integrated to generate a new multifunction product. A good deal of conceptual solutions of the new multifunction product can be acquired by analyzing, comparing, and integrating two sub function chains of the existing products under the help of the proposed approach. The proposed approach is effective for the conceptual solution generation of the mechanical and electrical multifunction products. But it is not known whether the approach is effective for other multifunction products, and the verification is needed in the future work. Meanwhile, the redesign process of the complex–correlation components is unordered, and mathematical tools are still needed.

References

1. Singh V, Walther B, Krager J et al (2007) Design for transformation: theory, method and application. In: Proceedings of the ASME international design engineering technical conferences and computers and information in engineering conference (IDETC/CIE'07), Las Vegas, Nev, September 2007, pp 447–459
2. Kuhr R, Wood K, Jensen D, Crawford R (2010) Concept opportunity diagrams, a visual modeling method to find multifunctional design concepts. In: Proceedings of the ASME

international design engineering technical conferences and computers and information in engineering conference (IDETC/CIE'10), Quebec, Canada, August 2010, pp 193–205
3. Feng PE, Zhang S, Pan SX, Chen Y, He B (2005) Cyclic solving process and realization for conceptual design of complex function product. Chin J Mech Eng 41(3):135–141
4. Hu GZ, Xiao SN, Liu ZB (2012) Fuzzy reconfigurable design system of complex mechanical products. Mod Manuf Eng 5:68–73
5. Li ZK (2009) Theories and methodologies for product family reconfigurable design and application in large scale air separation systems. D.E. thesis, Zhejiang University, Hangzhou
6. Zhang HG, Zhao WY, Jiang P, Tan RH (2012) Product platform design process model based on similarity and structural sensitivity analysis. J Mech Eng 48(11):104–118
7. Liu F, Jiang P, Zhang R-H, Tan R-H (2005) Platform design of function similar products based on bottom-up method. Comput Integr Manuf Syst 11(7):947–952
8. Wei Z, Tan JR, Feng YX, Feng X (2006) Construction of product module based on similarity and its applications. J Comput Aided Des Comput Gr 18(12):1930–1934
9. Weaver JM, Kuhr R, Wang D et al (2009) Increasing innovation in multi-function systems: Evaluation and experimentation of teo ideation methods for design. In: Proceedings of the ASME international design engineering technical conferences and computers and information in engineering conference IDETC/CIE '09), San Diego, Calif, September 2009, pp 965–983
10. Kang YY, Tang DB, Wang H (2012) Design solutions generation method based on function–component matrix and design structure matrix. Comput Integr Manuf Syst 18(5):897–904
11. Fomin P, Sarkani S, Mazzuchi TA (2010) Growing the house of quality. Syst Res Forum 4 (2):137–149
12. Kusiak A (1999) Engineering design: products, processes and systems. Academic Press, Orlando, FL
13. Tang DB, Zhang GJ, Dai S (2009) Design as integration of axiomatic design and design structure matrix. Robot Comput Integr Manuf 25(3):610–619
14. Ramanujan D, Bernstein W, Zhao F, Ramani K (2011) Addressing uncertainties within product redesign for sustainability: a function based framework. In: Proceedings of the ASME international design engineering technical conferences and computers and information in engineering conference (IDETC/CIE'11), Washington, DC, August 2011
15. Zhang JM, Wang JX, Wei XP (2008) Product modeling for conceptual design based on extended functional basis. Trans Chin Soc Agric Mach 39(1):129–133
16. Stone RB, Wood KL (2000) Development of a functional basis for design. J Mech Des 122 (4):359–370
17. Hirtz J, Stone RB, McAdams DA, Szykman S, Wood KL (2002) A functional basis for engineering design: reconciling and evolving previous efforts. Res Eng Des 13(2):65–82
18. Zou GM, Hu YJ, Xiao WS, Li CG (2004) Research on the model of product conceptual design based on the functional basis. Chin Mech Eng 15(3):206–210
19. McAdams D, Stone RB, Wood KL (1998) Understanding product similarity using customer needs. In: Proceedings of the ASME international design engineering technical conferences (IDETC'98), Atlanta, GA, September 1998
20. Qi GN, Gu XJ, Yang QH, Yu JH (2003) Principles and key technologies of mass customization. Comput Integr Manuf Syst 9(9):776–783

Chapter 4
Matrix-Based Computational Concept Design with Ant Colony Optimization

4.1 Introduction and Synopsis

Engineering design is a special form of problem solving where a set of frequently unclear objectives must be balanced without violating a set of constraints. As the concept generation phase of the engineering design becomes more complex and competitive, it is desirable to produce a large number of potential optimized solutions. However, completely exploring a problem's solution space is an unreasonable expectation for an unaided designer or design team. Designers mostly have limited options available for creating a quality design. Available options may include utilizing patent searches to find other approaches or similar designs, drawing on personal experiences or the experiences of co-workers, and reverse engineering existing products to evaluate how either the current design or a redesign could be used to meet the design goals [1]. These methods are potentially biased by a designer's experiences. Many methods and tools of conceptual design have been realized as computational algorithms to help designers [2–6]. However, these methods are potentially limited and tend to solve a special problem, such as analysis of consumer requirements [7] and solution optimization [2].

Quantification, formalization and automation of the conceptual design phase are active areas of research [8, 9]. Recently, computational tools have emerged to help designers more fully explore possible solutions. These automated concept generators use knowledge from existing designs and the desired functionality of the new product to suggest solutions [5, 10, 11]. These innovations indicate that larger design problems can be solved automatically. However, for all but the simplest products, these methods produce an unmanageably large set of solutions for the designer to evaluate. Techniques are needed to organize the concept solutions into smaller groups, parsed more easily by the human designer.

The work presented in this chapter shows the steps taken to automatically generate and evaluate design solutions based on a mathematical algorithm. A matrix-based computational method of concept generation that quickly produces

D. Tang et al., *Matrix-Based Product Design and Change Management*,
DOI 10.1007/978-981-10-5077-0_4

a manageable array of design solutions is proposed. The automatically generated concept solutions can then be used for product design directly or as a catalyst for generating additional design solutions through complimentary non-computational creative techniques. The specific focus of this research is the formalization of function-based synthesis, combinatorial optimization of design solutions and searching for feasible design candidates.

The following section presents a review of the related work establishing the foundation for our developments. In Sect. 4.3, the product representation and design tools used in this chapter are reviewed. In Sect. 4.4, computer-implemented techniques for automatically producing conceptual design solutions are then presented and illustrated effectively. In Sect. 4.5, a software prototype used in our creativity experiments is implemented, and a case study is introduced. Conclusions are presented and future work is directed in Sect. 4.6.

4.2 Related Research of Product Conceptual Design

We begin with a review of the state of the art in the area of conceptual design research and the area that supports automated concept generation. First, we review a functional model that is an abstraction for understanding and describing the design problem that is independent of specific possible solutions [12]. This work uses a centric device, operation on flows, to model. In short, the flows of material, energy and signal are introduced to describe the product and the sub-functions are operations on these flows. A product can be thought of as series of sub-functions selected and arranged to transform the available input flows into the desired output flows as specified by the overall function. Stone and Wood [13] and Hirtz et al. [14] developed a design language known as functional basis. It uses the function and flow words to form a sub-function description. Functional basis is intended to be broad enough to span the entire mechanical design space while not being repetitive.

A number of tools and techniques already exist to assist the designer during embodiment design. An initial attempt to assist the designer in the early stages of design is the design compiler [15]. The design complier took system schematics, specifications and utility functions as inputs and supplied a list of detailed components as outputs. This was effectively an automated catalogue design system. Unfortunately, this approach still required detailed knowledge of the form of the final design. Tools were still needed for the fuzzy front end of the problem [16].

Recently, two automated concept solution generators have been developed: one is a matrix-based method and the other is a graph grammar-based approach [17]. These two systems share a common set of inputs, but arrive at their results via different concept generation algorithms. Solutions from the graph-based approach have been sorted either using a penalty function-based search to select apparently optimal components [17], or using historical designer preferences for specific components [18]. The penalty function-based approach draws obvious parallels with other catalogue design approaches. However, it requires precise information that may not be present so early in the design. To support the design process, different matrix-based tools and methods have been developed. These matrices are typically used to describe and visualize the relationships between properties of the product and activities in the design process. One example of a matrix-based method is the House of Quality from quality function deployment (QFD), where customer requirements are mapped to engineering characteristics [19]. Other matrix-based methods include the incidence matrix [20], the design structure matrix (DSM) [21–25], the function impact matrix [3, 4] and the product relationship matrix [26]. Another matrix is the design matrix originating out of Axiomatic Design [27]. By applying different matrix operations, the design activities can be organized and analyzed. A third approach is similar to the matrix-based approach, and bridges some of the gap between these efforts and earlier design compliers [28]. However, this method requires far more detailed information about the final configuration.

The creators of the above concept generators acknowledge that the large sets of solutions produced by these concept generators are a significant hurdle for the designer [18, 29]. To alleviate some of the strain, Bryant allows the user to select one concept at a time and indicates the historical popularity of components [16]. However, this approach may decrease some of the advantage of the automated concept generator as it indulges the novice designer's desire to select only a few conceptual solutions. The designer suffers from the fact that designers' preferences may change from problem to problem.

Matrix-based methods above have been mostly combined with optimization methods to alleviate some of the strain. For example, customer satisfaction is optimized in QFD [30]. Incidence matrices are used together with optimization [20] and DSM is commonly used to structure multidisciplinary optimization problems [31–34]. And the design matrix from Axiomatic Design is also used to combine with optimization [24, 35]. An evaluation and optimization procedure for product life-cycle design at the conceptual stage is presented using a digraph and matrix approach [36]. Some intelligent algorithms were used in product optimization problem too, such as ant colony optimization (ACO) [2, 37], fuzzy set theory [38, 39] and neural network [40]. Recently, a new computational method is proposed to support product architecture and analyze the networks of components [41]. However, how to select the components according to the sub-function requirements is not introduced particularly in the chapter. Different from the methods above, this chapter presents an alternative conceptual design scheme that attempts to address some of these concerns based on matrix, functional basis and ACO.

4.3 Tools Used in Matrix-Based Conceptual Generation

The following sub-sections describe the design tools used for automation in the concept generation phase of the design process. First, the functional basis is presented as the means of both capturing the design knowledge and representing the conceptual design input required by the computational scheme. Second, the repository used to store the knowledge of the relationships between sub-functions and components is introduced. Third, the morphological matrix is described in detail. Finally, ACO for complex optimization is introduced.

4.3.1 Functional Basis of Design

The functional basis, where product function is characterized in a verb-object (function-flow) format, is a consistent language or coding system required to ensure that others can read it. It is intended to comprehensively describe the mechanical design space without repetition. The functional basis contains three primary flows and eight primary functions. The three primary flows are material, signal and energy. The material has five further specified secondary categories with an expanded list of tertiary categories. The signal class has two further specified secondary categories with an expanded list of tertiary categories. The energy has 13 further specified secondary categories with an expanded list of tertiary categories. The eight primary functions are branch, channel, connect, control magnitude, convert, provision, signal and support. Each primary function has several further specified secondary categories with an expanded list of tertiary categories. Clear definitions have been developed for all flows and functions [14]. Functional basis is applied to the areas of product architecture development, functional model generation, and design information archival and transmittal.

The functional model for any product can be generated with the functional basis. Generation of a black box model, creation of function chains for each input flow and aggregation of function chains are the sequential steps of a functional model construction [42]. The functional model also increases the clarity of the design problem by tracking of input and output flows. It is used to capture design knowledge from existing products and represent a form of independent blueprint of a product.

4.3.2 Function–Component Repository

Reverse engineering techniques are applied to existing consumer products [1], and the repository, where information extracted from product's bill of materials and the relationships between functions and components are stored, can be developed. A design repository from Stone et al. [43] is built over several years and includes currently 184 products and 6906 artefacts available for search. For freely handling the knowledge in our software prototype, we developed our own design repository including less-available products compared with Stone's repository. The knowledge stored in the repository is steadily expanding and benefits from a broad base of consumer products. Sub-functions are defined with the functional basis in the repository, and every sub-function is implemented by one or more components

Table 4.1 Contents stored in the repository

Components	Functions		Parameters		
	Function 1	Function 2	Economy	Noise	Torque
Linear motor	Import electrical energy	Convert electrical energy to rotational energy	0.8	0.5	0.8
DC motor	Import electrical energy	Convert electrical energy to rotational energy	0.8	0.3	0.8
AC motor	Import electrical energy	Convert electrical energy to rotational energy	0.5	0.3	0.4
Shaded pole induction motor	Import electrical energy	Convert electrical energy to rotational energy	0.4	0.4	0.9

stored in the repository. Components' parameters, such as economy, structure, noise and capacity, are stored in the repository too. Sub-functions and components are classified according to product categories. Each artefact is classified under a specific component name according to its function. For example, separate artefacts under different products may be named 'linear motor', 'shaded pole induction motor', 'dc motor' or 'AC motor', as given in Table 4.1. Using the functional basis naming scheme, each of these artefacts would be identified as a similar function and tagged as 'convert electrical energy to rotational energy'. Using this technique allows a well-defined function to be used in the creation of the morphological matrix and the function–component matrix (FCM) [3, 4], and helps in maintaining matrices of manageable size and eliminating artefact redundancies.

4.3.3 Function–Component Repository

Morphology is a method of thinking. One of the ideas of morphology is to systematically search for a solution to a problem by trying out all possible combinations in a matrix. The morphological matrix is created by decomposing the main function of the product into sub-functions which are listed on the vertical axis of the matrix. Possible solution principles for each function are then listed on the horizontal axis. Different concepts are created by combining various solution principles to form a complete system concept. The morphological matrix can be used to span all possible solutions to a problem.

An example of a morphological matrix for a motorcycle is given in Table 4.2. There are three functions defined by the functional basis: store electrical energy, convert electrical energy to rotational energy and stop rotational energy. Electrical energy could be stored in lead, NiCd or Li-ion batteries. There are four components that could convert electrical energy to rotational energy and three components that could stop rotational energy.

Table 4.2 A morphological matrix for a motorcycle

Functions	Components			
	Column 1	Column 2	Column 3	Column 4
Store electrical energy	Lead battery	NiCd battery	Li-ion battery	
Convert electrical energy to rotational energy	Linear motor	DC motor	AC motor	Shaded pole induction motor
Stop rotational energy	Disk brake	Drum brake	Regenerative electrical brake	

The major deficiency of the method is the large number of possible concepts that arise from already relatively small matrices. If there are m_1 possible components for function f_1 and m_2 components for function f_2 and so on, there are a total of $N = m_1 \times m_2 \times \cdots \times m_n$ possible concept combinations. The small matrix in Table 4.2 will give the designer 36 possible concept combinations. If there are many components and functions, the number of possible concept combinations is enormous. Another issue is that many components in a combination are not compatible with each other. Therefore, computerized optimization is necessary and a computerized optimization framework for the morphological matrix was applied to aircraft conceptual design [44].

4.3.4 Ant Colony Optimization

The study from biology of an ant colony shows that its behavior is highly structured. Inspired by this behavior, an algorithm named ACO was developed [45]. The basic principle of the algorithm is to have a population of artificial ants that cyclically construct a solution to a combinatorial optimization problem. In the solution space of a problem, each alternative for a solution is regarded as a node passed by artificial ants. The co-operating artificial ants communicate by means of a distributed memory signal named pheromone that is generated by an ant to share with other ants. The ants move along every branch from one node to another node according to the quantity of pheromone and so construct paths representing solutions. The algorithm consists of two rules: selection probability rule and pheromone update rule. Starting from the initial node, every ant chooses the next node in its path according to the rule as per Eq. (4.1):

$$p_{ij}(t) = \frac{\left[\tau_{ij}(t)\right]^{\alpha} \cdot \left[\frac{1}{d_{ij}}\right]^{\beta}}{\sum_{j \in allowed\ nodes} \left[\tau_{ij}(t)\right]^{\alpha} \cdot \left[\frac{1}{d_{ij}}\right]^{\beta}} \tag{4.1}$$

where τ_{ij} is the quantity of pheromone on the edge between node i and node j. d_{ij} is the heuristic distance between node i and node j. p_{ij} is the probability to branch from node i and node j [46].

The parameters α and β tune the relative importance in probability of the amount of pheromone versus the heuristic distance. Most likely, the definition of the problem will impose some restrictions on the sequence of nodes that generate valid solutions. According to this rule, nodes that have a higher amount of pheromone and that are closer to the actual node will have a higher probability to be chosen in the partial solution.

When all ants have constructed a complete solution, that is a sequence of visited nodes resulting in a complete solution to the problem, the cycle is complete and a pheromone update rule is applied according to Eqs. (4.2) and (4.3):

$$\tau_{ij}(t+n) = (1-\rho)\cdot\tau_{ij}(t) + \rho\cdot\Delta\tau_{ij}(t+n) \tag{4.2}$$

$$\Delta\tau_{ij}(t+n) = \begin{cases} \frac{Q}{f_{evaluation}(best\_so\_far)} \\ 0, \ otherwise \end{cases} \tag{4.3}$$

where ρ is the evaporation coefficient. $f_{evaluation}(best\_so\_far)$ is the length of the path. Q is the quantity of pheromone per unit of distance [46].

This rule consists of two actions. First, a fraction of pheromone on all edges is evaporated. Second, an increment of pheromone is given to those edges that are scheduled within the solution that has the best-so-far solution to the problem. In the next cycle, those edges belonging to the best-so-far solution will have higher probability; thus exploring this information performs a certain kind of reinforcement learning. On the other hand, evaporation prevents searching in the neighborhood of a local minimum [47].

4.4 Matrix-Based Concept Generation Method (MCGM)

Our proposed method utilizes the functional basis to describe sub-functions generated from the decomposition of the overall product function. The relationships between sub-functions and components, which describe compatibility of sub-functions and components from existing consumer products, are stored in the function–component repository. Thus, components compatible with sub-functions of a new product may be screened out from the repository and a morphological matrix will be structured correspondingly. Then, FCM and DSM that, respectively, describe the function–component relationship and the component–component compatibility of existing consumer products are generated from the morphological matrix [48]. Consequently, a large number of concept design solutions fit to functional requirement are generated by FCM and DSM computation. Finally, the

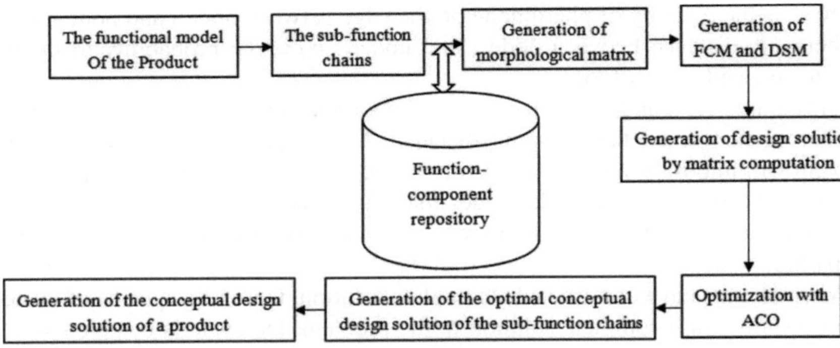

Fig. 4.1 Outline of the concept generation method

optimal conceptual design solution is filtered out by ACO according to a great variety of evaluation criteria. The steps of the concept generation scheme are summarized in Fig. 4.1.

The design knowledge stored in the repository, morphological matrix, FCM, DSM and ACO is used to generate, filter conceptual design solutions according to following sub-sections.

4.4.1 Function Decomposition and Generation of Sub-function Chain

The proposed concept generation scheme begins with function decomposition and functional model for either a new product to be developed or a previously developed product that is to be redesigned. We are in agreement as to the distinction between energy, material and signal, but it is possible that using an energy flow expresses the principal or primary flow associated with conducting the main function of the product. The material flow corresponds to an energy flow by unit of volume or mass. Similarly, a signal may be an expression of energy flow even though the energy level is relatively weak [49]. The main function and main energy flow resulting in the specific action are sought as a result of the product being designed. So we can decompose the overall function into several sub-function chains according to different types of transformation and transmission of energy. The main sub-function chain needed to structure the design functional model is generated according to main energy flow. Other sub-function chains can be generated by tracking other energy flows and functional model is structured to completely describe the design activity. Functional model are described by a set of functions and flows with precise definitions. This enables us to reconcile the design with physics based approaches in which input or output energy flows. This chapter presents an energy based view of functional decomposition.

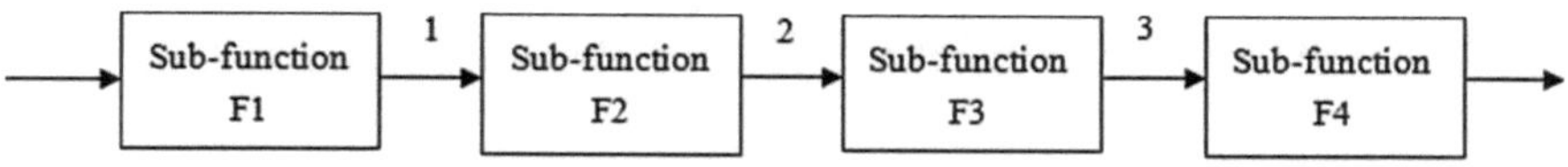

(a) Chain of function flow

	F1	F2	F4	F5
F1	0	1	0	0
F2	0	0	2	0
F3	0	0	0	3
F4	0	0	0	0

(b) Function-Function Matrix

Fig. 4.2 Sub-function chain

The function model of a product can be divided into several single non-branching sub-function chains [10, 13]. In the chapter, our concern is the solving method of design solution for a single non-branching sub-function chain. The design solutions of all single non-branching sub-function chains can be acquired in the same way. The design solution of a product can be acquired by combining design solutions of sub-function chains. For example, a non-branching sub-function chain is illustrated as shown in Fig. 4.2, where F1, F2, F3 and F4 are unspecified sub-functions of the product to be designed. The block diagram is translated into a matrix form that describes the connectivity between functions. Non-zero cell entries indicate forward connection between the row and column functions.

4.4.2 *Generation of FCM*

The components corresponding to sub-functions are searched out from the function–component repository. Morphological matrix is generated based on the searched knowledge and converted to FCM. The function–component relationships in the above theoretical example are shown in Fig. 4.3. Non-zero cell entries in the FCM indicate that components in the column can be applied to completely or partly implement the function in the row.

	1	2	3
F1	C1	C2	C3
F2	C4	C5	
F3	C5	C6	C7
F4	C7	C8	C9

(a) Morphological Matrix

	C1	C2	C3	C4	C5	C6	C7	C8	C9
F1	1	1	1	0	0	0	0	0	0
F2	0	0	0	1	1	0	0	0	0
F3	0	0	0	0	1	1	1	0	0
F4	0	0	0	0	0	0	1	1	1

(b) Function-Component Matrix

Fig. 4.3 FCM generated from the morphological matrix

4.4.3 Generation of Initial Solution Matrix

As shown in Fig. 4.4, the information from previous sections is used to generate ISM. First, the transpose of the row vector from the FCM that corresponds to each of the sub-functions from the flow chain in Sect. 4.4.1 is multiplied by the row vector from the FCM that corresponds to the forward connected sub-function, a new component–component matrix will be generated for each function connection in the flow chain [10]. Non-zero cells within these newly created component–component

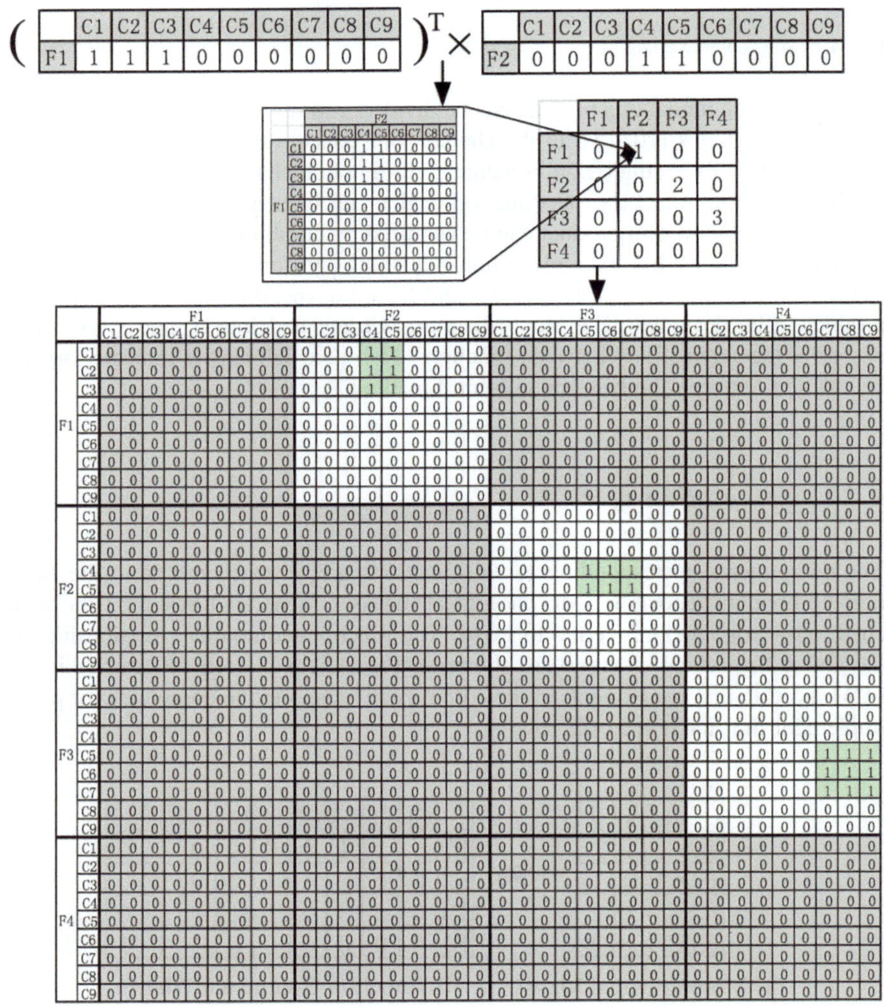

Fig. 4.4 Generation process of ISM

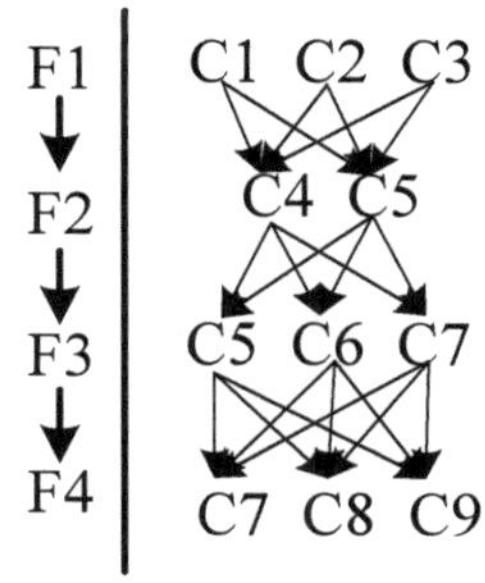

Fig. 4.5 Components tree

matrices represent all theoretically possible component combinations that will solve each pair of functions in the flow chain. Second, these component–component matrices are placed into the function–function matrix and then replace the cell entry "1". As a result, an ISM is generated. Tracing every possible "path" of connections in the solution matrix will acquire a list of all theoretically possible component chains that could potentially provide sub-function chain solutions. A component "tree" is created showing the chains of components, as shown in Fig. 4.5.

4.4.4 Definition of Component–Component Matrix

In this section, additional design knowledge gathered from existing consumer products is used to define the compatibility between components in the examined products. There are two manners for defining compatibility of components. One is using the relationship of components stored in repository. When we search components according to sub-function, the relationship of components can be searched out from the repository. The other is using designer's experiences to define. For instance, various measures of design needs (e.g. manufacturability, recyclability, failure etc.) entered as the non-zero FCM and/or DSM cell entries can be used to filter the resulting conceptual design solutions. The new relationship will be stored in repository to assist designer in the future. Meanwhile, the compatibility of components can be modified by designer for the special needs. Connection information for each artefact is stored in the database, and component–component matrices (frequently called DSM) for individual product or specified groups of products can easily be generated from these components' compatibility information. The compatibility information and DSM of the theoretical example are shown in Fig. 4.6. Non-zero cell entries in the DSM indicate that the component from the column has been directly connected to the component in the row.

Components	Compatible components with components in left cloumn
C1	C4, C5
C2	C4, C5
C3	C5
C4	C5, C6
C5	C6, C7, C8, C9
C6	C7, C9
C7	C7, C8
C8	
C9	

Design Structure Matrix

	C1	C2	C3	C4	C5	C6	C7	C8	C9
C1	0	0	0	1	1	0	0	0	0
C2	0	0	0	1	1	0	0	0	0
C3	0	0	0	0	1	0	0	0	0
C4	0	0	0	0	1	1	0	0	0
C5	0	0	0	0	0	1	1	1	1
C6	0	0	0	0	0	0	1	0	1
C7	0	0	0	0	0	0	1	1	0
C8	0	0	0	0	0	0	0	0	0
C9	0	0	0	0	0	0	0	0	0

Fig. 4.6 Generation of DSM

4.4.5 Filtration of Initial Conceptual Solution

The components compatibility information contained in the DSM is used to filter the ISM. As shown in Fig. 4.7, each component connection of each sub-function pair is checked for compatibility by the existing connection information in the DSM. Each cell of the DSM is multiplied with the corresponding cell of component–component matrices in each of the function pair of the solution matrix. Overlaying the DSM on each matrix of sub-function pair has the effect of removing any of the component connections that do not appear in DSM. This technique uses the "experience" contained in the DSM to filter out potentially inadequate concept solutions and reduce the set of possible concept solutions down to a more manageable size. After the matrices are filtered, we can once again trace every "path" of possible components to generate a list of feasible component chains that implement the function chain in Sect. 4.4.1.

Design solutions contained in filtered solution matrix can meet all requirements of function transfer and design structure. However, there are possibly a number of reasonable design solutions for further analysis and filtration by designer himself. This work is unpleasant and perplexed sometimes. So a concise and pleasant method for screening out the optimal design solution is necessary. A calculation method with evaluation value, evaluation criteria and weighting of sub-functions and criteria is proposed in the following sub-section.

4.4.6 Definition of Evaluation Criteria and Weighting

As shown in Fig. 4.8, evaluation criteria of components and weighting of both evaluation criteria and sub-functions were introduced. These are prerequisite for

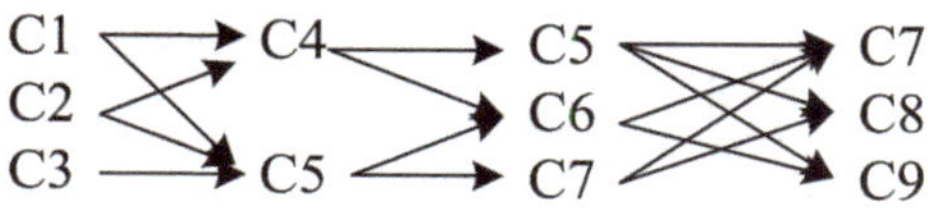

(a) Components tree pruned by incompatibility of components

		F1								
		C1	C2	C3	C4	C5	C6	C7	C8	C9
	C1	0	0	0	0	0	0	0	0	0
	C2	0	0	0	0	0	0	0	0	0
	C3	0	0	0	0	0	0	0	0	0
	C4	0	0	0	0	0	0	0	0	0
F1	C5	0	0	0	0	0	0	0	0	0
	C6	0	0	0	0	0	0	0	0	0
	C7	0	0	0	0	0	0	0	0	0
	C8	0	0	0	0	0	0	0	0	0
	C9	0	0	0	0	0	0	0	0	0
	C1	0	0	0	0	0	0	0	0	0
	C2	0	0	0	0	0	0	0	0	0
	C3	0	0	0	0	0	0	0	0	0
	C4	0	0	0	0	0	0	0	0	0
F2	C5	0	0	0	0	0	0	0	0	0
	C6	0	0	0	0	0	0	0	0	0
	C7	0	0	0	0	0	0	0	0	0
	C8	0	0	0	0	0	0	0	0	0
	C9	0	0	0	0	0	0	0	0	0
	C1	0	0	0	0	0	0	0	0	0
	C2	0	0	0	0	0	0	0	0	0
	C3	0	0	0	0	0	0	0	0	0
	C4	0	0	0	0	0	0	0	0	0
F3	C5	0	0	0	0	0	0	0	0	0
	C6	0	0	0	0	0	0	0	0	0
	C7	0	0	0	0	0	0	0	0	0
	C8	0	0	0	0	0	0	0	0	0
	C9	0	0	0	0	0	0	0	0	0
	C1	0	0	0	0	0	0	0	0	0
	C2	0	0	0	0	0	0	0	0	0
	C3	0	0	0	0	0	0	0	0	0
	C4	0	0	0	0	0	0	0	0	0
F4	C5	0	0	0	0	0	0	0	0	0
	C6	0	0	0	0	0	0	0	0	0
	C7	0	0	0	0	0	0	0	0	0
	C8	0	0	0	0	0	0	0	0	0
	C9	0	0	0	0	0	0	0	0	0

		F2								
		C1	C2	C3	C4	C5	C6	C7	C8	C9
	C1	0	0	0	1	1	0	0	0	0
	C2	0	0	0	1	1	0	0	0	0
	C3	0	0	0	0	1	0	0	0	0
	C4	0	0	0	0	0	0	0	0	0
F1	C5	0	0	0	0	0	0	0	0	0
	C6	0	0	0	0	0	0	0	0	0
	C7	0	0	0	0	0	0	0	0	0
	C8	0	0	0	0	0	0	0	0	0
	C9	0	0	0	0	0	0	0	0	0
	C1	0	0	0	0	0	0	0	0	0
	C2	0	0	0	0	0	0	0	0	0
	C3	0	0	0	0	0	0	0	0	0
	C4	0	0	0	0	0	0	0	0	0
F2	C5	0	0	0	0	0	0	0	0	0
	C6	0	0	0	0	0	0	0	0	0
	C7	0	0	0	0	0	0	0	0	0
	C8	0	0	0	0	0	0	0	0	0
	C9	0	0	0	0	0	0	0	0	0
	C1	0	0	0	0	0	0	0	0	0
	C2	0	0	0	0	0	0	0	0	0
	C3	0	0	0	0	0	0	0	0	0
	C4	0	0	0	0	0	0	0	0	0
F3	C5	0	0	0	0	0	0	0	0	0
	C6	0	0	0	0	0	0	0	0	0
	C7	0	0	0	0	0	0	0	0	0
	C8	0	0	0	0	0	0	0	0	0
	C9	0	0	0	0	0	0	0	0	0
	C1	0	0	0	0	0	0	0	0	0
	C2	0	0	0	0	0	0	0	0	0
	C3	0	0	0	0	0	0	0	0	0
	C4	0	0	0	0	0	0	0	0	0
F4	C5	0	0	0	0	0	0	0	0	0
	C6	0	0	0	0	0	0	0	0	0
	C7	0	0	0	0	0	0	0	0	0
	C8	0	0	0	0	0	0	0	0	0
	C9	0	0	0	0	0	0	0	0	0

		F3								
		C1	C2	C3	C4	C5	C6	C7	C8	C9
	C1	0	0	0	0	0	0	0	0	0
	C2	0	0	0	0	0	0	0	0	0
	C3	0	0	0	0	0	0	0	0	0
	C4	0	0	0	0	0	0	0	0	0
F1	C5	0	0	0	0	0	0	0	0	0
	C6	0	0	0	0	0	0	0	0	0
	C7	0	0	0	0	0	0	0	0	0
	C8	0	0	0	0	0	0	0	0	0
	C9	0	0	0	0	0	0	0	0	0
	C1	0	0	0	0	0	0	0	0	0
	C2	0	0	0	0	0	0	0	0	0
	C3	0	0	0	0	0	0	0	0	0
	C4	0	0	0	0	1	1	0	0	0
F2	C5	0	0	0	0	0	1	1	0	0
	C6	0	0	0	0	0	0	0	0	0
	C7	0	0	0	0	0	0	0	0	0
	C8	0	0	0	0	0	0	0	0	0
	C9	0	0	0	0	0	0	0	0	0
	C1	0	0	0	0	0	0	0	0	0
	C2	0	0	0	0	0	0	0	0	0
	C3	0	0	0	0	0	0	0	0	0
	C4	0	0	0	0	0	0	0	0	0
F3	C5	0	0	0	0	0	0	0	0	0
	C6	0	0	0	0	0	0	0	0	0
	C7	0	0	0	0	0	0	0	0	0
	C8	0	0	0	0	0	0	0	0	0
	C9	0	0	0	0	0	0	0	0	0
	C1	0	0	0	0	0	0	0	0	0
	C2	0	0	0	0	0	0	0	0	0
	C3	0	0	0	0	0	0	0	0	0
	C4	0	0	0	0	0	0	0	0	0
F4	C5	0	0	0	0	0	0	0	0	0
	C6	0	0	0	0	0	0	0	0	0
	C7	0	0	0	0	0	0	0	0	0
	C8	0	0	0	0	0	0	0	0	0
	C9	0	0	0	0	0	0	0	0	0

		F4								
		C1	C2	C3	C4	C5	C6	C7	C8	C9
	C1	0	0	0	0	0	0	0	0	0
	C2	0	0	0	0	0	0	0	0	0
	C3	0	0	0	0	0	0	0	0	0
	C4	0	0	0	0	0	0	0	0	0
F1	C5	0	0	0	0	0	0	0	0	0
	C6	0	0	0	0	0	0	0	0	0
	C7	0	0	0	0	0	0	0	0	0
	C8	0	0	0	0	0	0	0	0	0
	C9	0	0	0	0	0	0	0	0	0
	C1	0	0	0	0	0	0	0	0	0
	C2	0	0	0	0	0	0	0	0	0
	C3	0	0	0	0	0	0	0	0	0
	C4	0	0	0	0	0	0	0	0	0
F2	C5	0	0	0	0	0	0	0	0	0
	C6	0	0	0	0	0	0	0	0	0
	C7	0	0	0	0	0	0	0	0	0
	C8	0	0	0	0	0	0	0	0	0
	C9	0	0	0	0	0	0	0	0	0
	C1	0	0	0	0	0	0	0	0	0
	C2	0	0	0	0	0	0	0	0	0
	C3	0	0	0	0	0	0	0	0	0
	C4	0	0	0	0	0	0	0	0	0
F3	C5	0	0	0	0	0	0	1	1	1
	C6	0	0	0	0	0	0	1	0	1
	C7	0	0	0	0	0	0	1	1	0
	C8	0	0	0	0	0	0	0	0	0
	C9	0	0	0	0	0	0	0	0	0
	C1	0	0	0	0	0	0	0	0	0
	C2	0	0	0	0	0	0	0	0	0
	C3	0	0	0	0	0	0	0	0	0
	C4	0	0	0	0	0	0	0	0	0
F4	C5	0	0	0	0	0	0	0	0	0
	C6	0	0	0	0	0	0	0	0	0
	C7	0	0	0	0	0	0	0	0	0
	C8	0	0	0	0	0	0	0	0	0
	C9	0	0	0	0	0	0	0	0	0

(b) Solution matrix equivalent with component tree

Fig. 4.7 Two forms of reasonable design solution

automatically screening out optimal design solutions. As for different design solutions, the importance of each sub-function decomposed from the overall function is different, and the weighting of sub-function is used to describe the different importance. The sum of weightings of all sub-functions is 1. Each sub-function has one or more evaluation criteria and each criterion has different weighting for relevant sub-function, and the sum of weightings of all evaluation criteria for a sub-function is 1. Meanwhile, each sub-function is implemented by one or more components. For each evaluation criterion of a sub-function, there are different evaluation values for different components. These values and weightings of sub-function or evaluation criterion are stored in function–component repository, and can be defined, added, modified or deleted according to designer's experience

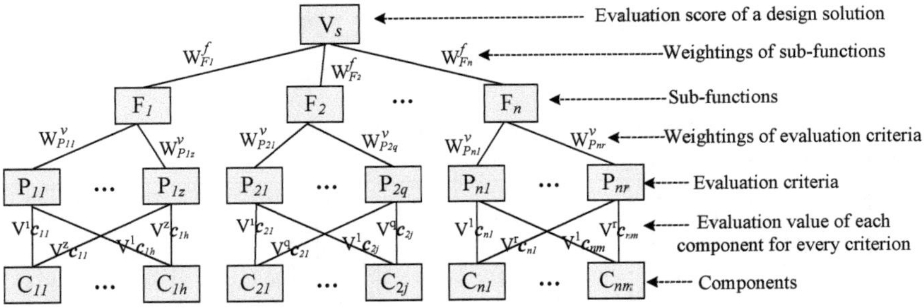

Fig. 4.8 Tree of evaluation

or special requirement of new product. The designer's creative ability and new product's specific character may be embodied by defining these values and weightings. Computationally, the sum of component's evaluation value multiplied by weighting of criterion is calculated out firstly, and next be multiplied by the weighting of sub-function. At last, the evaluation score of a design solution is calculated out by adding the evaluation score of each sub-function together. For example, there is a design solution X and it was composed by a component chain $(C_{11}, C_{2j}, \ldots, C_{nm})$. Its evaluation score can be calculated by following Eq. (4.4):

$$\begin{aligned} V_s^X = & (V_{C_{11}}^1 \cdot W_{P_{11}}^v + V_{C_{11}}^2 \cdot W_{P_{12}}^v + \cdots + V_{C_{11}}^z \cdot W_{P_{1z}}^v) \cdot W_{F1}^f \\ & + (V_{C_{2j}}^1 \cdot W_{P_{21}}^v + V_{C_{2j}}^2 \cdot W_{P_{22}}^v + \cdots + V_{C_{2j}}^q \cdot W_{P_{2q}}^v) \cdot W_{F2}^f \\ & + \cdots + (V_{C_{nm}}^1 \cdot W_{P_{n1}}^v + V_{C_{nm}}^2 \cdot W_{P_{n1}}^v + \cdots + V_{C_{nm}}^r \cdot W_{P_{nr}}^v) \cdot W_{Fn}^f \end{aligned} \tag{4.4}$$

where X represents a design solution. Different sub-function may have several same criteria, and its weighting may be different. Criteria used to evaluate each component and evaluation value for relevant criterion may be different. The following section introduces a method for automatically calculating and optimizing design solution.

4.4.7 *Optimization of Design Solution with ACO*

In this section, ACO is used to optimize and select the best design solutions. ACO has been applied to many combinatorial optimization problems, ranging from quadratic assignment to protein folding or routing vehicles [50]. A lot of derived methods have been adapted to dynamic problems in real variables [51], stochastic problems, multi-targets and parallel implementations. It has also been used to produce near-optimal solutions to the travelling salesman problem. They have an advantage over simulated annealing and genetic algorithm approaches of similar problems when the graph may change dynamically. The ACO can be run

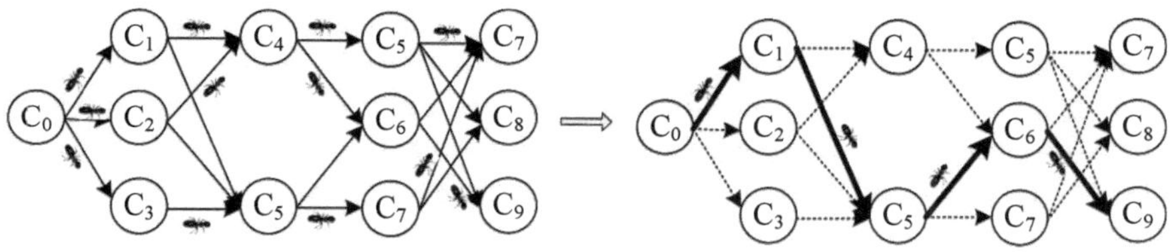

Fig. 4.9 Optimization process with ACO

continuously and adapt to changes in real time [52]. Thus in this chapter, ACO is applied with presented matrices. The optimization process with ACO can be described by a graph shown in Fig. 4.9. The graph is structured according to the components tree shown in Fig. 4.7a. Components in the tree are described as nodes passed by ants and compatible relationships of components are described as path. Meanwhile, node C_0 is regarded as the initial site of all ants as shown in Fig. 4.9. Using the graph, the problem of selecting optimal design solution is converted to a problem of seeking shortest path. The ACO is used to seek shortest path of the graph. Every ant selects own path freely according to established rule composed by set of equations and at last all ants will track a same path to end node. The final path composed by components is the optimal result corresponding to the optimal design solution. The optimization process is illustrated in Fig. 4.9.

Regarding Sect. 4.4.6, the total evaluation value of all components (nodes) passed by an ant from initial node to end node is calculated out by Eq. (4.5):

$$V_k = \sum_{i=0}^{n} w_i^f \sum_{\mu=1}^{z} w_\mu^v \varphi(e_\mu) + \delta \sum_{i=0}^{n} \eta_{ij}(t) \tag{4.5}$$

where parameter V_k is the score of the evaluation value of a design solution composed by components (nodes) passed by an ant and equation $\sum_{i=0}^{n} w_i^f \sum_{\mu=1}^{z} w_\mu^v \varphi(e_\mu)$ is another representation of V_s^X in Eq. (4.4). The parameter n is the total number of sub-functions. The parameter i and μ is variable quantity representing respectively sub-function and evaluation criterion. w_i^f and w_μ^v represent the weighting of sub-function and evaluation criterion respectively, $\sum_{i=0}^{n} w_i^f = 1$ and $\sum_{\mu=1}^{z} w_\mu^v = 1$. The parameter z is the total number of evaluation criteria. The parameter e_μ represents an evaluation criterion. Function $\varphi(e_\mu)$ is used to calculate evaluation value of a criterion corresponding to different components. For instance, $V_{C_{11}}^1$ is a result of $\varphi(e_\mu)$. Sum of evaluation value of one component for i sub-function is calculated by equation $\sum_{\mu=1}^{z} w_\mu^v \varphi(e_\mu)$. For ACO, $\eta_{ij}(t)$ is heuristic function and its value is $\frac{1}{d_{ij}}$, and the parameter d_{ij} is the length of path (i, j) [45]. But in Eq. (4.5), the value of $\eta_{ij}(t)$ is r_{ij}. The parameter r_{ij} represents the compatibility of component i and component j. If the compatibility of component i and j is greater, the selection probability of branch (*i*, *j*) is greater. The parameter δ is a balance coefficient and its value is defined according to designer's experience.

When all ants have constructed a complete solution, that is a sequence of visited nodes resulting in a complete solution to the problem, a cycle is completed and pheromone update rules are applied, which is shown in Eqs. (4.6)–(4.8):

$$\Delta\tau_{ij}^{k}(t) = \begin{cases} Q \cdot V_k, & \text{if ant k pass through the path (i,j) in a cycle} \\ 0, & \text{otherwise} \end{cases} \tag{4.6}$$

$$\Delta\tau_{ij}(t) = \sum_{k=1}^{m} \Delta\tau_{ij}^{k}(t) \tag{4.7}$$

$$\tau_{ij}(t+1) = (1-\rho)\tau_{ij}(t) + \Delta\tau_{ij}(t) \tag{4.8}$$

where parameter Q is convergence speed coefficient influenced by pheromone intensity. The parameter m is the total number of ants. Function $\Delta\tau_{ij}(t)$ represents the increment of pheromone influenced by all ants passing through the branch (*i*, *j*) in one cycle. Function $\tau_{ij}(t)$ represents pheromone residues on the branch (*i*, *j*) before a cycle and function $\tau_{ij}(t+1)$ represents pheromone residues on the branch (*i*, *j*) after the cycle. The parameter ρ is volatile coefficient and the range of its value is from 0 to 1.

Using the algorithm, the optimization problem is defined into the graph, in which the ants move along every branch from one node to another and so construct paths representing solutions. Starting in a node *i*, every ant chooses the next node *j* in its path according to the state transition rule shown in Eq. (4.9):

$$P_{ij}^{k}(t) = \begin{cases} \frac{[\tau_{ij}(t)]^{\alpha} \cdot [\eta_{ij}(t)]^{\beta}}{\sum_{s \in W} [\tau_{is}(t)]^{\alpha} \cdot [\eta_{is}(t)]^{\beta}}, & \text{if } j \in W = allowed_k \\ 0, & \text{otherwise} \end{cases} \tag{4.9}$$

where $P_{ij}^{k}(t)$ represents the selection probability of a component on the node *j* for ant *k* on the node *i*. The parameter α is information stimulating factor. It represents the relative importance of pheromone and reflects the effect of pheromone accumulated in ants moving process when an ant chooses next objective. If its value is greater, collaboration between ants is stronger, and an ant is inclined to pass after other ants. The parameter β is expectation stimulating factor. It represents the relative importance of objective visibility and reflects the effect of expected factor when an ant chooses next objective. If its value is greater, selection probability is closer to greed rules. The parameter *t* represents cycle index. It is one cycle that all ants move from the initial node to the end node. The parameter *W* indicates the collection of alternative components in next node *j*. As shown in Fig. 4.10, the relationship of *i*, *j* and *W* is illustrated and the value of *W* is $\{j_1, j_3\}$.

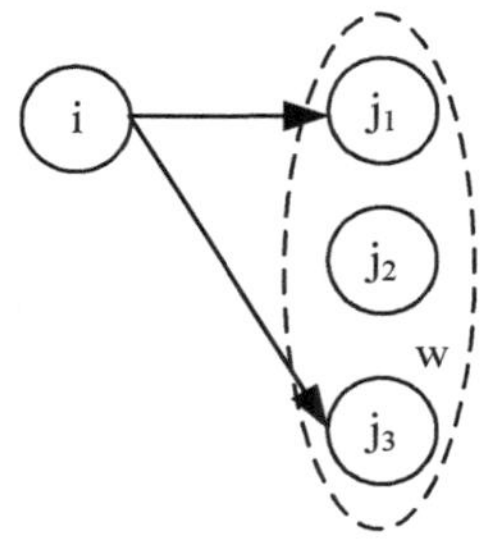

Fig. 4.10 Relations of i, j and W

The rules introduced above are repeatedly used by artificial ants to search the optimal solution. If there are lots of design solutions, the number of cycle is enormous. The cycle is terminated according to the following equation:

$$[V^g(t) - V^g(t-1)] \leq \varepsilon \tag{4.10}$$

where parameter $V^g(t)$ is the highest evaluation score of the last cycle and parameter $V^g(t-1)$ is the highest evaluation score of the cycle before last one. The parameter ε is termination condition and its value is given by executives while the optimization program is executed. And the detailed process of optimization program execution is introduced as following steps:

Step 1: Initializing parameter

The initial value of cycle index t is 0. The quantity of ants is m, and all ants are placed at node C_0, as shown in Fig. 4.9. Initial value of other parameters are set as follows: $\Delta\tau_{ij}(0) = 0$, $\alpha = 1$, $\beta = 1$, $V^g(0) = 0$, $\delta = 0$. Incompatible connection for components was filtered out by DSM in Sect. 4.4.5. The similarity coefficient r_{ij} of all connected components is postulated equal and the value of r_{ij} is 1, so the initial value of heuristic function $\eta_{ij}(0)$ is 1 too.

When the optimization program is executing, pheromone of each branch and selection probability are recorded in matrix. Initial pheromone $(\tau_{ij}(0))$ matrix is generated from design solution matrix, as shown in Fig. 4.11. A virtual sub-function F_0 and a virtual component C_0 are added in the matrix. Pheromone on the branch from C_0 to C_1, C_2 and C_3 is equal. Thus selection probability of C_1, C_2 and C_3 for an ant is equal too. The ants only choose the path on which pheromone persists. In the pheromone matrix and probability matrix, components in the row are regarded as the start of each branch, and components in the column are the end of each branch. Sub-functions (F_0 to F_4) in the row indicate the node passed by ant colony.

As shown in Fig. 4.12, initial probability $(P_{ij}^k(0))$ matrix comes from initial pheromone matrix according to the rule defined in Eq. (4.9). In the first cycle, all ants choose the branch according to the initial probability in matrix.

		F1									F2									F3									F4								
		C1	C2	C3	C4	C5	C6	C7	C8	C9	C1	C2	C3	C4	C5	C6	C7	C8	C9	C1	C2	C3	C4	C5	C6	C7	C8	C9	C1	C2	C3	C4	C5	C6	C7	C8	C9
F0	C0	1	1	1	0	0	0	0	0	0	0	0	0	0	0	0	0	0	0	0	0	0	0	0	0	0	0	0	0	0	0	0	0	0	0	0	0
	C0	0	0	0	0	0	0	0	0	0	0	0	0	0	0	0	0	0	0	0	0	0	0	0	0	0	0	0	0	0	0	0	0	0	0	0	0
	C0	0	0	0	0	0	0	0	0	0	0	0	0	0	0	0	0	0	0	0	0	0	0	0	0	0	0	0	0	0	0	0	0	0	0	0	0
	C0	0	0	0	0	0	0	0	0	0	0	0	0	0	0	0	0	0	0	0	0	0	0	0	0	0	0	0	0	0	0	0	0	0	0	0	0
	C0	0	0	0	0	0	0	0	0	0	0	0	0	0	0	0	0	0	0	0	0	0	0	0	0	0	0	0	0	0	0	0	0	0	0	0	0
	C0	0	0	0	0	0	0	0	0	0	0	0	0	0	0	0	0	0	0	0	0	0	0	0	0	0	0	0	0	0	0	0	0	0	0	0	0
	C0	0	0	0	0	0	0	0	0	0	0	0	0	0	0	0	0	0	0	0	0	0	0	0	0	0	0	0	0	0	0	0	0	0	0	0	0
	C0	0	0	0	0	0	0	0	0	0	0	0	0	0	0	0	0	0	0	0	0	0	0	0	0	0	0	0	0	0	0	0	0	0	0	0	0
	C0	0	0	0	0	0	0	0	0	0	0	0	0	0	0	0	0	0	0	0	0	0	0	0	0	0	0	0	0	0	0	0	0	0	0	0	0
F1	C1	0	0	0	0	0	0	0	0	0	0	0	0	1	1	0	0	0	0	0	0	0	0	0	0	0	0	0	0	0	0	0	0	0	0	0	0
	C2	0	0	0	0	0	0	0	0	0	0	0	0	1	1	0	0	0	0	0	0	0	0	0	0	0	0	0	0	0	0	0	0	0	0	0	0
	C3	0	0	0	0	0	0	0	0	0	0	0	0	0	1	0	0	0	0	0	0	0	0	0	0	0	0	0	0	0	0	0	0	0	0	0	0
	C4	0	0	0	0	0	0	0	0	0	0	0	0	0	0	0	0	0	0	0	0	0	0	0	0	0	0	0	0	0	0	0	0	0	0	0	0
	C5	0	0	0	0	0	0	0	0	0	0	0	0	0	0	0	0	0	0	0	0	0	0	0	0	0	0	0	0	0	0	0	0	0	0	0	0
	C6	0	0	0	0	0	0	0	0	0	0	0	0	0	0	0	0	0	0	0	0	0	0	0	0	0	0	0	0	0	0	0	0	0	0	0	0
	C7	0	0	0	0	0	0	0	0	0	0	0	0	0	0	0	0	0	0	0	0	0	0	0	0	0	0	0	0	0	0	0	0	0	0	0	0
	C8	0	0	0	0	0	0	0	0	0	0	0	0	0	0	0	0	0	0	0	0	0	0	0	0	0	0	0	0	0	0	0	0	0	0	0	0
	C9	0	0	0	0	0	0	0	0	0	0	0	0	0	0	0	0	0	0	0	0	0	0	0	0	0	0	0	0	0	0	0	0	0	0	0	0
F2	C1	0	0	0	0	0	0	0	0	0	0	0	0	0	0	0	0	0	0	0	0	0	0	0	0	0	0	0	0	0	0	0	0	0	0	0	0
	C2	0	0	0	0	0	0	0	0	0	0	0	0	0	0	0	0	0	0	0	0	0	0	0	0	0	0	0	0	0	0	0	0	0	0	0	0
	C3	0	0	0	0	0	0	0	0	0	0	0	0	0	0	0	0	0	0	0	0	0	0	0	0	0	0	0	0	0	0	0	0	0	0	0	0
	C4	0	0	0	0	0	0	0	0	0	0	0	0	0	0	0	0	0	0	0	0	0	0	1	1	0	0	0	0	0	0	0	0	0	0	0	0
	C5	0	0	0	0	0	0	0	0	0	0	0	0	0	0	0	0	0	0	0	0	0	0	0	1	1	0	0	0	0	0	0	0	0	0	0	0
	C6	0	0	0	0	0	0	0	0	0	0	0	0	0	0	0	0	0	0	0	0	0	0	0	0	0	0	0	0	0	0	0	0	0	0	0	0
	C7	0	0	0	0	0	0	0	0	0	0	0	0	0	0	0	0	0	0	0	0	0	0	0	0	0	0	0	0	0	0	0	0	0	0	0	0
	C8	0	0	0	0	0	0	0	0	0	0	0	0	0	0	0	0	0	0	0	0	0	0	0	0	0	0	0	0	0	0	0	0	0	0	0	0
	C9	0	0	0	0	0	0	0	0	0	0	0	0	0	0	0	0	0	0	0	0	0	0	0	0	0	0	0	0	0	0	0	0	0	0	0	0
F3	C1	0	0	0	0	0	0	0	0	0	0	0	0	0	0	0	0	0	0	0	0	0	0	0	0	0	0	0	0	0	0	0	0	0	0	0	0
	C2	0	0	0	0	0	0	0	0	0	0	0	0	0	0	0	0	0	0	0	0	0	0	0	0	0	0	0	0	0	0	0	0	0	0	0	0
	C3	0	0	0	0	0	0	0	0	0	0	0	0	0	0	0	0	0	0	0	0	0	0	0	0	0	0	0	0	0	0	0	0	0	0	0	0
	C4	0	0	0	0	0	0	0	0	0	0	0	0	0	0	0	0	0	0	0	0	0	0	0	0	0	0	0	0	0	0	0	0	0	0	0	0
	C5	0	0	0	0	0	0	0	0	0	0	0	0	0	0	0	0	0	0	0	0	0	0	0	0	0	0	0	0	0	0	0	0	0	1	1	1
	C6	0	0	0	0	0	0	0	0	0	0	0	0	0	0	0	0	0	0	0	0	0	0	0	0	0	0	0	0	0	0	0	0	0	1	0	1
	C7	0	0	0	0	0	0	0	0	0	0	0	0	0	0	0	0	0	0	0	0	0	0	0	0	0	0	0	0	0	0	0	0	0	1	1	0
	C8	0	0	0	0	0	0	0	0	0	0	0	0	0	0	0	0	0	0	0	0	0	0	0	0	0	0	0	0	0	0	0	0	0	0	0	0
	C9	0	0	0	0	0	0	0	0	0	0	0	0	0	0	0	0	0	0	0	0	0	0	0	0	0	0	0	0	0	0	0	0	0	0	0	0
F4	C1	0	0	0	0	0	0	0	0	0	0	0	0	0	0	0	0	0	0	0	0	0	0	0	0	0	0	0	0	0	0	0	0	0	0	0	0
	C2	0	0	0	0	0	0	0	0	0	0	0	0	0	0	0	0	0	0	0	0	0	0	0	0	0	0	0	0	0	0	0	0	0	0	0	0
	C3	0	0	0	0	0	0	0	0	0	0	0	0	0	0	0	0	0	0	0	0	0	0	0	0	0	0	0	0	0	0	0	0	0	0	0	0
	C4	0	0	0	0	0	0	0	0	0	0	0	0	0	0	0	0	0	0	0	0	0	0	0	0	0	0	0	0	0	0	0	0	0	0	0	0
	C5	0	0	0	0	0	0	0	0	0	0	0	0	0	0	0	0	0	0	0	0	0	0	0	0	0	0	0	0	0	0	0	0	0	0	0	0
	C6	0	0	0	0	0	0	0	0	0	0	0	0	0	0	0	0	0	0	0	0	0	0	0	0	0	0	0	0	0	0	0	0	0	0	0	0
	C7	0	0	0	0	0	0	0	0	0	0	0	0	0	0	0	0	0	0	0	0	0	0	0	0	0	0	0	0	0	0	0	0	0	0	0	0
	C8	0	0	0	0	0	0	0	0	0	0	0	0	0	0	0	0	0	0	0	0	0	0	0	0	0	0	0	0	0	0	0	0	0	0	0	0
	C9	0	0	0	0	0	0	0	0	0	0	0	0	0	0	0	0	0	0	0	0	0	0	0	0	0	0	0	0	0	0	0	0	0	0	0	0

Fig. 4.11 Matrix of initial pheromone

Step 2: Iteration process of the optimization program

While not $[V^g(t) - V^g(t-1)] \leq \varepsilon$ do

For $k = 1$ to m (m is the population of ants),

For $i = 0$ to n (n is the total number of sub-functions),

An ant k chooses path based on the probability in the probability matrix. The value of V_k for the route passed by ant k is calculated out according to Eq. (4.5).

End for i.

The value of V_k is recorded and compared with the value of V_{k-1}, then the smaller is removed.

End for k.

		F1									F2									F3									F4								
		C1	C2	C3	C4	C5	C6	C7	C8	C9	C1	C2	C3	C4	C5	C6	C7	C8	C9	C1	C2	C3	C4	C5	C6	C7	C8	C9	C1	C2	C3	C4	C5	C6	C7	C8	C9
	C0	1/3	1/3	1/3	0	0	0	0	0	0	0	0	0	0	0	0	0	0	0	0	0	0	0	0	0	0	0	0	0	0	0	0	0	0	0	0	0
	C0	0	0	0	0	0	0	0	0	0	0	0	0	0	0	0	0	0	0	0	0	0	0	0	0	0	0	0	0	0	0	0	0	0	0	0	0
	C0	0	0	0	0	0	0	0	0	0	0	0	0	0	0	0	0	0	0	0	0	0	0	0	0	0	0	0	0	0	0	0	0	0	0	0	0
	C0	0	0	0	0	0	0	0	0	0	0	0	0	0	0	0	0	0	0	0	0	0	0	0	0	0	0	0	0	0	0	0	0	0	0	0	0
F0	C0	0	0	0	0	0	0	0	0	0	0	0	0	0	0	0	0	0	0	0	0	0	0	0	0	0	0	0	0	0	0	0	0	0	0	0	0
	C0	0	0	0	0	0	0	0	0	0	0	0	0	0	0	0	0	0	0	0	0	0	0	0	0	0	0	0	0	0	0	0	0	0	0	0	0
	C0	0	0	0	0	0	0	0	0	0	0	0	0	0	0	0	0	0	0	0	0	0	0	0	0	0	0	0	0	0	0	0	0	0	0	0	0
	C0	0	0	0	0	0	0	0	0	0	0	0	0	0	0	0	0	0	0	0	0	0	0	0	0	0	0	0	0	0	0	0	0	0	0	0	0
	C0	0	0	0	0	0	0	0	0	0	0	0	0	0	0	0	0	0	0	0	0	0	0	0	0	0	0	0	0	0	0	0	0	0	0	0	0
	C1	0	0	0	0	0	0	0	0	0	0	0	0	1/2	1/2	0	0	0	0	0	0	0	0	0	0	0	0	0	0	0	0	0	0	0	0	0	0
	C2	0	0	0	0	0	0	0	0	0	0	0	0	1/2	1/2	0	0	0	0	0	0	0	0	0	0	0	0	0	0	0	0	0	0	0	0	0	0
	C3	0	0	0	0	0	0	0	0	0	0	0	0	0	1	0	0	0	0	0	0	0	0	0	0	0	0	0	0	0	0	0	0	0	0	0	0
	C4	0	0	0	0	0	0	0	0	0	0	0	0	0	0	0	0	0	0	0	0	0	0	0	0	0	0	0	0	0	0	0	0	0	0	0	0
F1	C5	0	0	0	0	0	0	0	0	0	0	0	0	0	0	0	0	0	0	0	0	0	0	0	0	0	0	0	0	0	0	0	0	0	0	0	0
	C6	0	0	0	0	0	0	0	0	0	0	0	0	0	0	0	0	0	0	0	0	0	0	0	0	0	0	0	0	0	0	0	0	0	0	0	0
	C7	0	0	0	0	0	0	0	0	0	0	0	0	0	0	0	0	0	0	0	0	0	0	0	0	0	0	0	0	0	0	0	0	0	0	0	0
	C8	0	0	0	0	0	0	0	0	0	0	0	0	0	0	0	0	0	0	0	0	0	0	0	0	0	0	0	0	0	0	0	0	0	0	0	0
	C9	0	0	0	0	0	0	0	0	0	0	0	0	0	0	0	0	0	0	0	0	0	0	0	0	0	0	0	0	0	0	0	0	0	0	0	0
	C1	0	0	0	0	0	0	0	0	0	0	0	0	0	0	0	0	0	0	0	0	0	0	0	0	0	0	0	0	0	0	0	0	0	0	0	0
	C2	0	0	0	0	0	0	0	0	0	0	0	0	0	0	0	0	0	0	0	0	0	0	0	0	0	0	0	0	0	0	0	0	0	0	0	0
	C3	0	0	0	0	0	0	0	0	0	0	0	0	0	0	0	0	0	0	0	0	0	0	0	0	0	0	0	0	0	0	0	0	0	0	0	0
	C4	0	0	0	0	0	0	0	0	0	0	0	0	0	0	0	0	0	0	0	0	0	0	1/2	1/2	0	0	0	0	0	0	0	0	0	0	0	0
F2	C5	0	0	0	0	0	0	0	0	0	0	0	0	0	0	0	0	0	0	0	0	0	0	0	1/2	1/2	0	0	0	0	0	0	0	0	0	0	0
	C6	0	0	0	0	0	0	0	0	0	0	0	0	0	0	0	0	0	0	0	0	0	0	0	0	0	0	0	0	0	0	0	0	0	0	0	0
	C7	0	0	0	0	0	0	0	0	0	0	0	0	0	0	0	0	0	0	0	0	0	0	0	0	0	0	0	0	0	0	0	0	0	0	0	0
	C8	0	0	0	0	0	0	0	0	0	0	0	0	0	0	0	0	0	0	0	0	0	0	0	0	0	0	0	0	0	0	0	0	0	0	0	0
	C9	0	0	0	0	0	0	0	0	0	0	0	0	0	0	0	0	0	0	0	0	0	0	0	0	0	0	0	0	0	0	0	0	0	0	0	0
	C1	0	0	0	0	0	0	0	0	0	0	0	0	0	0	0	0	0	0	0	0	0	0	0	0	0	0	0	0	0	0	0	0	0	0	0	0
	C2	0	0	0	0	0	0	0	0	0	0	0	0	0	0	0	0	0	0	0	0	0	0	0	0	0	0	0	0	0	0	0	0	0	0	0	0
	C3	0	0	0	0	0	0	0	0	0	0	0	0	0	0	0	0	0	0	0	0	0	0	0	0	0	0	0	0	0	0	0	0	0	0	0	0
	C4	0	0	0	0	0	0	0	0	0	0	0	0	0	0	0	0	0	0	0	0	0	0	0	0	0	0	0	0	0	0	0	0	0	0	0	0
F3	C5	0	0	0	0	0	0	0	0	0	0	0	0	0	0	0	0	0	0	0	0	0	0	0	0	0	0	0	0	0	0	0	0	0	1/3	1/3	1/3
	C6	0	0	0	0	0	0	0	0	0	0	0	0	0	0	0	0	0	0	0	0	0	0	0	0	0	0	0	0	0	0	0	0	0	1/2	0	1/2
	C7	0	0	0	0	0	0	0	0	0	0	0	0	0	0	0	0	0	0	0	0	0	0	0	0	0	0	0	0	0	0	0	0	0	1/2	1/2	0
	C8	0	0	0	0	0	0	0	0	0	0	0	0	0	0	0	0	0	0	0	0	0	0	0	0	0	0	0	0	0	0	0	0	0	0	0	0
	C9	0	0	0	0	0	0	0	0	0	0	0	0	0	0	0	0	0	0	0	0	0	0	0	0	0	0	0	0	0	0	0	0	0	0	0	0
	C1	0	0	0	0	0	0	0	0	0	0	0	0	0	0	0	0	0	0	0	0	0	0	0	0	0	0	0	0	0	0	0	0	0	0	0	0
	C2	0	0	0	0	0	0	0	0	0	0	0	0	0	0	0	0	0	0	0	0	0	0	0	0	0	0	0	0	0	0	0	0	0	0	0	0
	C3	0	0	0	0	0	0	0	0	0	0	0	0	0	0	0	0	0	0	0	0	0	0	0	0	0	0	0	0	0	0	0	0	0	0	0	0
	C4	0	0	0	0	0	0	0	0	0	0	0	0	0	0	0	0	0	0	0	0	0	0	0	0	0	0	0	0	0	0	0	0	0	0	0	0
F4	C5	0	0	0	0	0	0	0	0	0	0	0	0	0	0	0	0	0	0	0	0	0	0	0	0	0	0	0	0	0	0	0	0	0	0	0	0
	C6	0	0	0	0	0	0	0	0	0	0	0	0	0	0	0	0	0	0	0	0	0	0	0	0	0	0	0	0	0	0	0	0	0	0	0	0
	C7	0	0	0	0	0	0	0	0	0	0	0	0	0	0	0	0	0	0	0	0	0	0	0	0	0	0	0	0	0	0	0	0	0	0	0	0
	C8	0	0	0	0	0	0	0	0	0	0	0	0	0	0	0	0	0	0	0	0	0	0	0	0	0	0	0	0	0	0	0	0	0	0	0	0
	C9	0	0	0	0	0	0	0	0	0	0	0	0	0	0	0	0	0	0	0	0	0	0	0	0	0	0	0	0	0	0	0	0	0	0	0	0

Fig. 4.12 Matrix of initial probability

The optimal evaluation score $V^g(t)$ is the highest score of V_1 to V_k and the best route is screened out in this cycle. Pheromone $\tau_{ij}(t)$ on each branch is renewed according to the rule. The pheromone matrix is renewed meanwhile. The probability matrix is renewed according to Eq. (4.9) correspondingly.

End for a while.

Step 3: The optimal evaluation score and the best route of each cycle are compared with each other, and the highest evaluation score and the best route of all cycles are acquired.

Step 4: End of program.

The flow chart of the above design solution optimization with ACO is shown in Fig. 4.13.

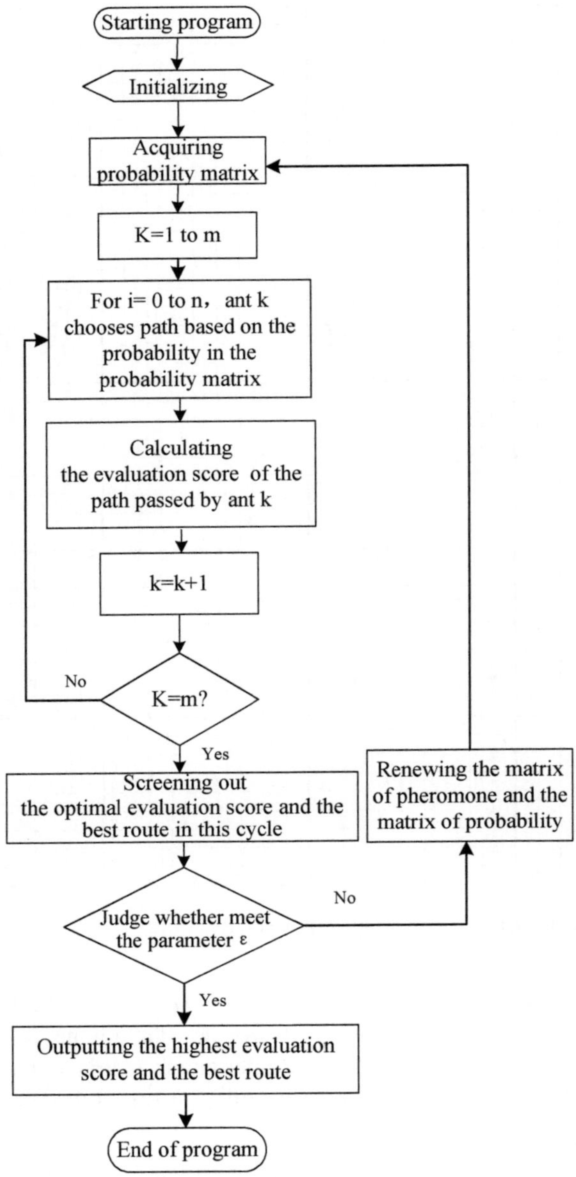

Fig. 4.13 Flow chart of design solution optimization

4.5 Application of MCGM for Product Design

With the design solution generation technology introduced in Sect. 4.4, a software prototype was developed to automatically produce concept design solutions for a sub-function chain. The prototype includes ten modules and function of every module is introduced in Sect. 4.4. For a new product, the overall function is decomposed to several sub-function chains in the first module. Sub-functions are defined by unified verb and noun words of the functional basis selected from the repository of existing products. Function model are generated according to the sub-function chains. Morphological matrix and FCM of a sub-function chain are generated according to sub-functions and artificial components selected automatically from the repository. Next, initial theoretical solution matrix is calculated out in creating solution matrix module. The relationships of artificial components in the FCM are defined by designer and DSM is generated in the DSM module. Feasible design solutions meeting requirement of function and structure are generated by filtering theoretical solution with DSM. Finally, evaluation criteria are defined and design solutions are optimized in the last three modules, so optimized solution of a sub-function chain is generated.

A case study is presented to demonstrate how the automated concept generation software effectively assists a designer during the early phases of application design. A design team was charged with redesigning transmission mechanism of a material loading platform. Transmission distance of existing device is shorter and can't meet the application requirement. It is necessary to redesign the loading platform. Transmission mechanism is the critical unit of the loading platform. It is required that the transmission components of existing device are used as far as possible. So under the assistance of the developed software prototype, a satisfactory design solution of new product is generated. Example of transmission mechanism is introduced to describe the process of solution generation. The main flow transferred by the device is energy flow, so the influence of energy flow is mostly analyzed but material flow and signal flow are rarely considered in the design activity. After determining the customer needs of the device to be designed, the transmission mechanism's functional model including one sub-function chain is generated, as shown in Fig. 4.14. The functional model is described with the functional basis words selected out from repository with the assist of the developed software.

Seven sub-functions were obtained from the functional model, and they are *import electrical energy*, *convert electrical energy to rotational energy*, *convert rotational energy to translational energy*, *convert translational energy to rotational energy*, *transmit rotational energy*, *transmit translational energy* and *export translational energy*. Next, artificial components were selected from the repository on the basis of sub-functions. As a result, function–function matrix, morphological matrix, FCM and DSM are generated automatically under the software operations by designer. For example, the generation process of FCM is shown in Fig. 4.15. First, the sequence number of sub-function chain was inputted and button *go* was clicked. Sub-functions were displayed in *sub-functions* column. Second, designer

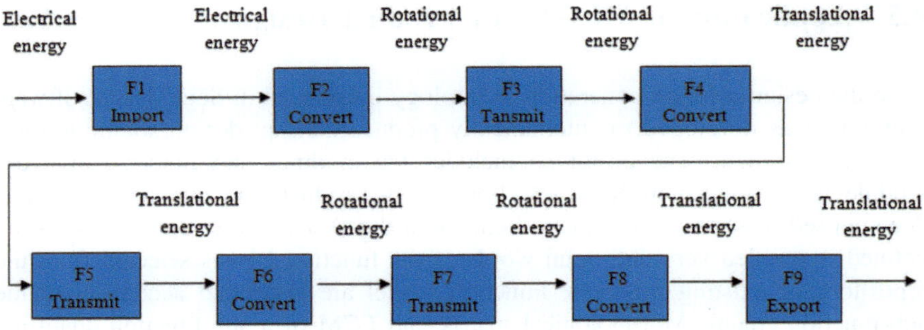

Fig. 4.14 Functional model of the transmission mechanism

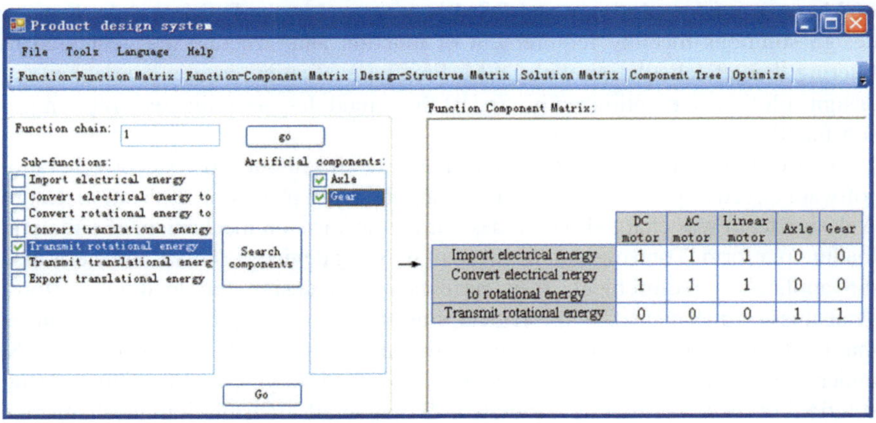

Fig. 4.15 Generation interface of FCM

selects a sub-function and click button *search components*, and corresponding components are displayed in *artificial components* column. Finally, designer selects components and clicks button *go*, a FCM is generated and displayed in *Function Component Matrix* column. When components were selected out from the repository, some components were used to construct FCM, and some components were discarded according to designer's experience or requirement of new product. Innovation of designer may be embodied.

After feasible solution matrix is generated, evaluation criteria and weightings of criterion are defined by designer on the basis of experience and existing device. Evaluation criteria and values of first three sub-functions for this case are shown in Table 4.3.

With the evaluation values and weightings, Ant Colony Optimization is used to search appropriate design solutions. Calculation process and initial values of parameters of ACO are shown in Fig. 4.16.

Table 4.3 Evaluation criteria of components for sub-functions

Function sequence	Sub-function	Weighting of function (w_i^f)	Evaluation criteria, weighting of criterion and evaluation values of components							
F1	Import electrical energy	0.05	Evaluation criteria	Economy	Structure	Noise	Torque	Speed adjustment mode	Power source	Overload capacity
			Weighting of criteria ($w_\mu^v \times 10$)	1.5	1	2	2	1	1	1.5
			DC motor	0.8	0.6	0.5	0.8	0.6	0.6	0.7
			AC motor	0.8	0.6	0.3	0.8	0.8	0.7	0.5
			Linear motor	0.5	0.6	0.3	0.4	0.8	0.7	0.5
F2	Convert electrical energy to rotational energy	0.1	Evaluation criteria	Economy	Structure	Noise	Torque	Speed adjustment mode	Power source	Overload capacity
			Weighting of criteria ($w_\mu^v \times 10$)	1.5	1	2	2	1	1	1.5
			DC motor	0.8	0.6	0.5	0.8	0.6	0.6	0.7
			AC motor	0.8	0.6	0.3	0.8	0.8	0.7	0.5
			Linear motor	0.5	0.6	0.3	0.4	0.8	0.7	0.5
F3	Transmit rotational energy	0.1	Evaluation criteria	Economy	Structure	Noise	Reusability	Transmission accuracy	Stationary	
			Weighting of criteria ($w_\mu^v \times 10$)	3	1	1	1	2	2	
			Axle	0.8	0.6	0.9	0.8	0.1	0.8	
			Gear	0.5	0.6	0.5	0.8	0.9	0.8	

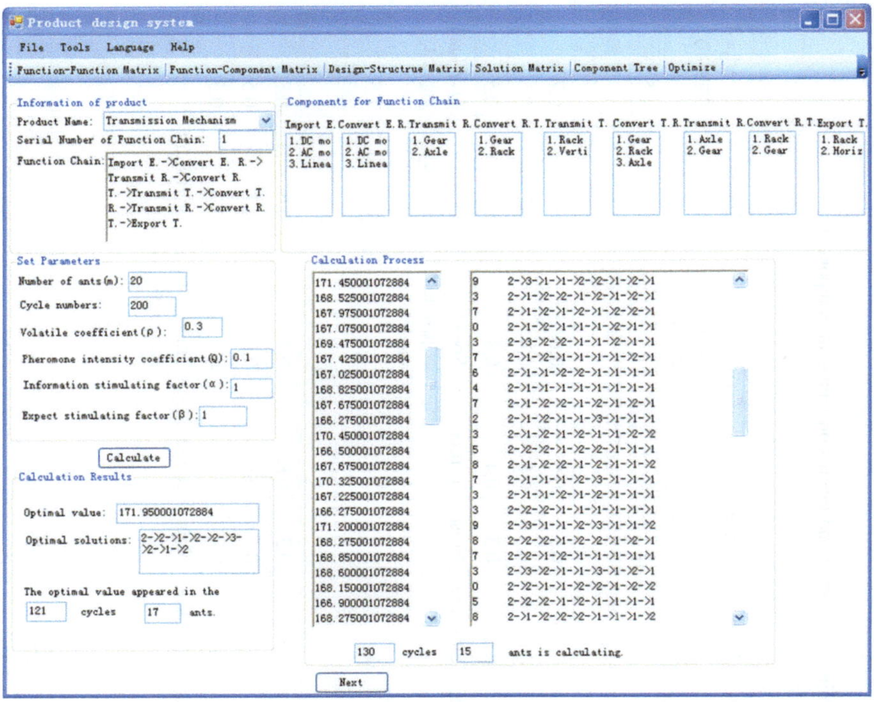

Fig. 4.16 Calculation interface with ACO

Component chains were generated during the course of program execution. The optimal component chain is displayed firstly and others then are ranked according to decreasing score. These component chains were displayed for designer to gather ideas. The ranked results are demonstrated as shown in Fig. 4.17.

After component chains are generated, evaluated and ranked, the designer would then be able to browse through these component chains to confirm suitable candidate solution or gather ideas or search more detailed information for a specific design requirement. In this case, the evaluation score of the first component chain is higher than other chains' evaluation scores. So the first component chain was ranked in the first place and the design solution in the first chain is the suitable candidate solution according to evaluation criteria. Then it was used to produce a new product, as shown in Fig. 4.18. Although the eventual embodiment of the device may be modified during the later stages of design from the initial conceptual solution presented in the first component chain, the concept solution may show the catalyzed idea that leads to a successful end design.

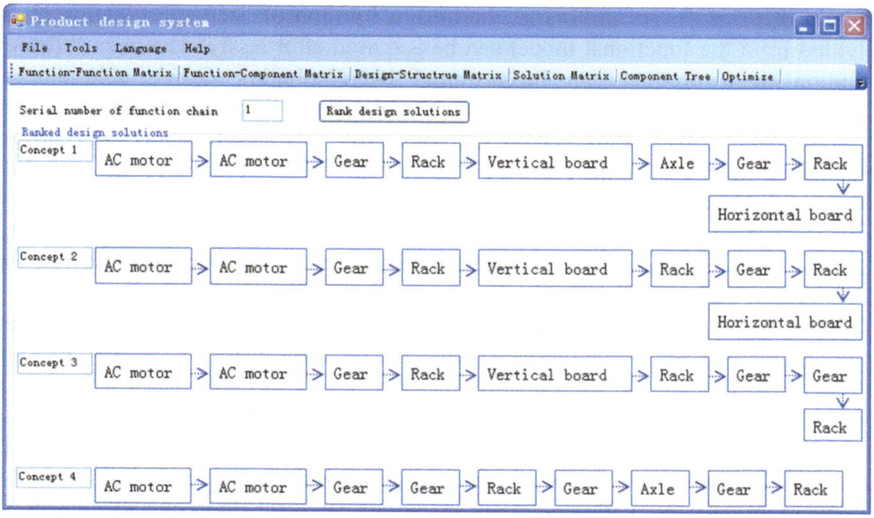

Fig. 4.17 Ranked component chains according to evaluation score

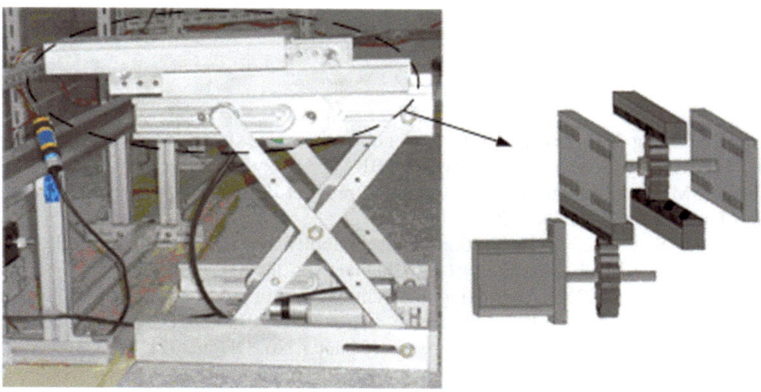

Fig. 4.18 Embodied design of the transmission mechanism

4.6 Conclusion and Future Work

A matrix-based concept generation and optimization method is introduced in this chapter. Intended to help designers choose suitable components for given functions in a redesign or original design, the goal of the method is to utilize existing design knowledge to rapidly produce a large array of reasonable candidate solutions in the early design process. This approach is typically more appropriate for reverse engineering applications when the artefact is being analyzed. In the chapter, a lot of effort was taken to introduce the method by analyzing a simple example. But with the aid of the software prototype, the design solution of each sub-function chain divided from the functional model can be acquired effortlessly. The overall design solution of a product can be acquired by combining design solutions of sub-function chains. The current software prototype is limited to handling single, non-branching sub-function chain automatically. The combination isn't implemented automatically in the current version of the software, and it is accomplished by designer's analysis. For a more complex product, the combination of multiple sub-optimums is necessary. Generally, the main design scheme of a product can be confirmed according to the design solution of the main sub-function chain of the main energy flow.

Next, we will investigate automatic calculation and optimization for complex energy, material and signal flow chains. Future improvements of the software prototype may enhance the capability to generate concept solutions for a full functional model. This may require evaluation criteria and weightings for a whole component chain. Furthermore, a particular cost-benefit analysis of the method will be addressed, and the implemented technique needs to be evaluated for efficiency about large matrices. Although basic in its current form, the concept generation technique presented has great potential for development into a powerful design tool. Current research to further this work is currently underway.

References

1. Tang DB, Kang YY, Zhu RM (2011) Functional reverse design for secondary innovation. Int J Prod Lifecycle Manag 5(2-3-4):183–201
2. Albritton MD, McMullen PR (2007) Optimal product design using a colony of virtual ants. Eur J Oper Res 176(1):498–520
3. Devanathan S et al (2010) Integration of sustainability into early design through the function impact matrix. J Mech Des 132(8):1–8 (Chapter No. 081004)
4. Ramanujan D et al (2011) Addressing uncertainties within product redesign for sustainability: a function based framework. In: Proceedings of ASME 2011 IDETC/CIE conference, 28–31 Aug 2011, Washington, DC, USA, pp 1–8 (Chapter No. DETC2011-47137)
5. Tang DB et al (2010) Product design knowledge management based on design structure matrix. Adv Eng Inf 24(2):159–166
6. Wang Z, Kannan PK, Azarm S (2011) Customer-driven optimal design for convergence products. J Mech Des 133(10):1–13 (Chapter No. 101010)
7. Kamyab A, Lewis K E (2008) Customizing products using functional component matrices. In: Proceedings of ASME 2008 IDETC/CIE conference, 3–6 Aug 2008, Brooklyn, New York, USA, pp 1–11 (Chapter No. DETC2008-50005)
8. Akasaka F et al (2012) Development of a knowledge-based design support system for product-service systems. Comput Ind 63(4):309–318
9. Tang DB et al (2010) Analysis of engineering change impacts based on design structure matrix. J Mech Eng 46(1):154–161
10. Bryant C et al (2005) A computational technique for concept generation. In: Proceedings of ASME 2005 IDETC/CIE conference, 24–28 Sept 2005, Long Beach, CA, pp 1–10 (Chapter No. DETC2005-85323)
11. Luh DB, Ko YT, Ma DH (2011) A structural matrix-based modeling for designing product variety. J Eng Des 22(1):1–29

12. Pahl G, Beitz W (1984) Engineering design. The Design Council, London
13. Stone R, Wood K (2000) Development of a functional basis for design. J Mech Des 122 (4):359–370
14. Hirtz J et al (2002) A functional basis for engineering design: reconciling and evolving previous efforts. Res Eng Des 13(2):65–82
15. Ward AC (1989) A theory of quantitative inference for artifact sets, applied to a mechanical design compiler. Thesis (PhD), Massachusetts Institute of Technology
16. Hyman B (1998) Fundamentals of engineering design. Prenice-Hall, Saddle River, NJ
17. Tamhankar M, Campbell M (2007) An intelligent and efficient tree search algorithm for computer-aided component selection. In: Proceedings of ASME 2007 IDETC/CIE conference, 4–7 Sept 2007, Las Vegas, NV, pp 1–10 (Chapter No. DETC2007-34587)
18. Kurtoglu T et al (2005) Capturing empirically derived design knowledge for creating conceptual design configurations. In: Proceedings of ASME 2005 IDETC/CIE conference, 24–28 Sept 2005, Long Beach, CA, pp 1–9 (Chapter No. DETC2005-84405)
19. Fomin P, Sarkani S, Mazzuchi T (2010) Growing the house of quality. Syst Res Forum 4 (2):137–149
20. Kusiak A (1999) Engineering design: products, processes and systems. Academic Press, Orlando, FL
21. Avnet MS, Weigel AL (2010) An application of the design structure matrix to integrated concurrent engineering. Acta Astronaut 66(5–6):937–949
22. Eppinger S et al (1994) A model-based method for organizing tasks in product development. Res Eng Des 6(1):1–13
23. Helmer R, Yassine A, Meier C (2010) Systematic module and interface definition using component design structure matrix. J Eng Des 21(6):647–675
24. Tang DB, Zhang GJ, Dai S (2009) Design as integration of axiomatic design and design structure matrix. Robot Comput Integr Manuf 25(3):610–619
25. Tilstra AH, Seepersad CC, Wood KL (2009) Analysis of product flexibility for future evolution based on design guidelines and a high-definition design structure matrix. In: Proceedings of ASME 2009 IDETC/CIE conference, 30 Aug–2 Sept 2009, San Diego, California, pp 1–14 (Chapter No. DETC2009-87118)
26. Seuring S (2009) The product-relationship-matrix as framework for strategic supply chain design based on operations theory. Int J Prod Econ 120(1):221–232
27. Suh N (2001) Axiomatic design advances and applications. Oxford University Press, New York
28. Tiwari S, Summers J, Fadel G (2007) A genetic algorithm based procedure for extracting optimal solutions from a morphological chart. In: Proceedings of ASME 2007 IDETC/CIE conference, 4–7 Sept 2007, Las Vegas, NV, pp 1–10 (Chapter No. DETC2007-35497)
29. Bryant C et al (2006) A validation study of an automated concept generator design tool. In: Proceedings of ASME 2006 IDETC/CIE conference, 10–13 Sept 2006, Philadelphia, PA, pp 1–12 (Chapter No. DETC2006-99489)
30. Lai X, Xie M, Tan KC (2006) QFD optimization using linear physical programming. Eng Optim 38(5):593–607
31. Braun RD (1996) Collaborative optimization: an architecture for large-scale distributed design. Thesis (Ph.D), Stanford University
32. Michelena N et al (1999) Hierarchical overlapping coordination for large-scale optimization by decomposition. AIAA J 37(7):890–906
33. Shi XM et al (2010) Time evaluation of design mission based on design structure matrix. In: Proceedings of international conference on information networking and automation (ICINA), vol 1, 18–19 Oct 2010, pp 470–474
34. Sobieszczanski-Sobieski J, Haftka RT (1997) Multidisciplinary aerospace design optimization: survey of recent developments. Struct Optim 14(1):1–23
35. Thielman J, Ge P (2006) Applying axiomatic design theory to the evaluation and optimization of large-scale engineering systems. J Eng Des 17(1):1–16

36. Anand A, Wani MF (2010) Product life-cycle modeling and evaluation at the conceptual design stage: a digraph and matrix approach. J Mech Des 132(9):1–9 (Chapter No. 091010)
37. Wagner MO et al (2003) Ergonomic modeling and optimization of the keyboard arrangement with an ant colony algorithm. J Eng Des 14(2):187–208
38. Ayağ Z (2005) A fuzzy AHP-based simulation approach to concept evaluation in a NPD environment. IIE Trans 37(9):827–842
39. Vanegas LV, Labib AW (2005) Fuzzy approaches to evaluation in engineering design. J Mech Des 127(1):24–33
40. Huang HZ, Bo RF, Chen W (2006) An integrated computational intelligence approach to product concept generation and evaluation. Mech Mach Theory 41(5):567–583
41. Wyatt DF et al (2012) Supporting product architecture design using computational design synthesis with network structure constraints. Res Eng Des 23(1):17–52
42. Kurfman M et al (2001) Functional modeling experimental studies. In: Proceedings of ASME 2001 DETC conference, 9–12 Sept 2001, Pittsburgh, PA, pp 1–13 (Chapter No. DETC2001/DTM-21709)
43. Stone R et al (2006) Design repository [online]. The Design Engineering Lab at Oregon State University, Corvallis, OR. Available from: http://function2.mime.oregonstate.edu:8080/view/index.jsp. Accessed 23 Nov 2012
44. Ölvander J, Lundén B, Gavel H (2009) A computerized optimization framework for the morphological matrix applied to aircraft conceptual design. Comput Aided Des 41(3):187–196
45. Dorigo M, Gambardella LM (1997) Ant colony system: a cooperative learning approach to the travelling salesman problem. IEEE Trans Evol Comput 1(1):53–66
46. van der Zwaan S, Marques C (1999) Ant colony optimization for job shop scheduling. In: Proceedings of the 3rd workshop on genetic algorithms and artificial life (GAAL99), March, Lisbon, Portugal, pp 1–8
47. Minseok S, José AV (2008) Ant colony optimization for job shop scheduling. In: IIE annual conference and expo, Vancouver, BC, pp 1433–1438
48. Malmqvist J, Axelsson R, Johansson M (1996) A comparative analysis of the theory of inventive problem solving and the systematic approach of Pahl and Beitz. In: Proceedings of ASME 1996 DETC conference, 18–22 Aug 1996, Irvine, CA, pp 1–11 (Chapter No. DETC/DTM-1529)
49. Pailhès J et al (2011) Energy based functional decomposition in preliminary design. J Mech Des 133(5):1–10 (Chapter No. 051011)
50. Fox B, Xiang W, Lee HP (2007) Industrial applications of the ant colony optimization algorithm. Int J Adv Manuf Technol 31(7–8):805–814
51. Zhou R, Nee AYC, Lee HP (2009) Performance of an ant colony optimization algorithm in dynamic job shop scheduling problems. Int J Prod Res 47(11):2903–2920
52. Selvi V, Umarani D (2010) Comparative analysis of ant colony and particle swarm optimization techniques. Int J Comput Appl 5(4):1–6

Chapter 5
Workload-Based Change Propagation Analysis in Product Design

5.1 Introduction and Synopsis

Engineering change (EC) is a significant activity in industry. EC may occur from concept, through definition, and development to manufacture, and then into service [1]. Comprehensive product management is the key to business enterprise management and change management is the core of product management. In terms of change management, a key measure is the engineering change impact analysis in order to conduct computational analysis and forecast evaluation. Since an engineering product is composed of numerous components and subsystems through mutual relationships among them, changes originating from a component or subsystem will propagate within the product, which may further complicate the impact analysis of the ECs. However, the actual design and production situation of enterprises shows that the activities of corporate engineering change management are mainly focused on tracking and recording engineering changes and lack quantitative analysis and evaluation of engineering changes [2].

Due to many tasks in product design process, design changes often occur, which can propagate through the process and generate significant impacts on the product development cycle [3]. High workload would normally lead to high risk, high cost, and long development time, etc. If the designer can predict the change propagation paths, the workload, and find the optimal change solution based on the change propagation analysis results, the design change can be implemented efficiently, so that the quality of the design change can also be assured. Requirements play a critical role within any design process and are maintained throughout the process [4–9]. Most design projects tend to evolve as the design process is an iterative process and, as a result, requirements will do the same, which will lead to design changes. In this chapter, requirement is related to several components and each of the components can be an originating change component to satisfy the change requirement. An originating change component corresponds to different change propagation paths. This chapter develops a method for searching the optimal

D. Tang et al., *Matrix-Based Product Design and Change Management*,
DOI 10.1007/978-981-10-5077-0_5

change propagation path for a requirement driven change, which includes: searching the change propagation paths on the component level, calculating the workload of design change, and further selecting the optimal change propagation path with the least workload. Besides, the method uses a matrix-based approach and two types of change propagation pattern (AND, OR) to generate change propagation paths.

The structure of the chapter is organized as follows. In Sect. 5.2, review of related research work is given. Section 5.3 introduces the general outline of the method and Sect. 5.4 puts forward how to search change propagation paths originating from a requirement. A detailed description of calculating the workload and analyzing the change propagation is presented in Sect. 5.5. Subsequently, by using a blowing and suction machine as an example, Sect. 5.6 demonstrates how the method can be applied. In Sects. 5.7 and 5.8, the method is initially evaluated and discussed respectively. A conclusion is given in Sect. 5.9.

5.2 Related Research of Engineering Change

Different reports suggest that ECs use around one-third of the engineering design capacity [10–12]. Thus EC management is crucial in the product design. It can comprise the five guidelines from Fricke et al. [11]: "less", "earlier", "more effective", "more efficient", and "better". These guidelines suggest that: first, the occurrence of changes should be avoided as far as possible; second, changes should be detected as early as possible to reduce their impact; third, changes should be selected more effectively; fourth, changes should be implemented more efficiently. Finally, the organization should learn from the past changes to continuously improve the management of engineering changes.

In the studies of ECs, the change items can be organizations, processes, product components, parameters, etc. Based on the STEP (Standard for the Exchange of Product) data model [13–15], Cohen et al. [16] develop a methodology called Change Favorable Representation (C-FAR), using existing product parameters, e.g., bottle size and bottle material, to facilitate change representation, propagation, and qualitative evaluation. Jan et al. [17] use characteristics to describe the shape and the structure of a product, e.g., geometry, materials, and properties to describe the behavior of a product, e.g., weight, manufacturability, function, cost. A change propagation path is a solution for the change requirement and the change propagation path is composed of properties and corresponding characteristics. A large data set from industry, including 41,500 change requests, is analyzed by Giffin et al. [18]. In the change networks, three motifs composed of components are introduced by Giffin et al. [18] to represent the building blocks, i.e., 1-, 2- and 3-change motifs. Yang and Duan [19] and Cohen et al. [16] apply parameters of components to study the change propagation and impact assessment. Yang and Duan [19] search change propagation paths based on a product model from the parameter linkage perspective. Unfortunately, the method requires far more detailed information about the product model.

Purposes of studying ECs can be diverse. Clarkson et al. [20] analyze direct and indirect changes to predict the risk of change propagation in terms of likelihoods and impacts of changes. Edwin et al. [21] assess the changeability of engineering systems in order to identify suitable components to be made more changeable. Oh et al. [22] analyze the change propagation and solve mathematical problems in order to design change-absorbing system architecture. Li and Zhao [23] propose a process simulation based method to select the most economic propagation path for each design change. However, the efficient solution of design change in terms of workload is rarely considered in the literature.

Besides, a variety of techniques have been developed to study ECs. As a structured modeling method, design structure matrix (DSM) has been widely used to record relationships between items. DSM is a good tool mapping information flow and recording its impact in product [1]. Tang et al. [1] apply three-domain DSMs, i.e., DSM of product, organization, process, to manage the change process. Relationships between components are formatted in the DSM of a product and items in the DSM contain the dependencies between components, e.g., high level material interaction and low level physical interaction. Likelihoods and impacts of changes between system components are recorded in the DSM to assess the changeability of complex engineering systems [20, 21]. Clarkson et al. [20] analyze the change behavior of complex design based on a case study in Westland Helicopters of rotorcraft design. In the study, change impacts and risks are computed with mathematical models and change likelihood is acted as a weight of change impact. Edwin et al. [21] provide another mathematical model to analyze the changeability of engineering systems. Through combining simulation of change propagation with optimization algorithms, Li and Zhao [23] present an integrated approach for scheduling design changes in the complex product development process. However, the generation process of change propagation paths is effort-consuming.

Logic relationship exists in the product at both functional and structural levels [24], and it can directly lead to the design process, which converts the functional needs into a technical solution, consisting of complicated logic dependencies between design tasks [23]. In the process simulation based method for scheduling product design change propagation, the AND/OR relationship is used to represent the input and output logics of each design change task by Li et al. [23]. In their research, the logic relationship is determined in terms of the duration of design change tasks. In our work, logic relationship between components is applied to search the change propagation paths more effectively. Different from the method above, the logic relationship is judged through component parameters.

To conclude, first, parameters are rarely considered for the judgment of logic relationship between components in current researches. Second, how to evaluate the schemes of design change is still an open issue and analysis of change propagation in terms of workload is not paid much attention to in the literature. In this chapter, workload is expressed as the consumption of resources and development time caused by the requirement change and change propagation. High workload normally leads to high consumption of resources and development time. Thus low workload is preferable to promote efficient implementation of design change. Smith and Eppinger [25] propound the concept of work transformation matrix (WTM), which records the strength of dependency between tasks. The strength of dependency between tasks is measured by the percentage of workload created for a task by work performed by other interdependent tasks. By referring to Smith and Eppinger [25], the concept of workload transformation is applied and a mathematical model is propounded to quantitatively measure the impacts between components in terms of workload. According to the workload of different change propagation paths, the change propagation can be analyzed and the scheme of design change can be implemented efficiently.

5.3 Outline of the Method

EC can originate from customer needs, reliability requirements, cost reduction, etc., which are referred as change requirements in this chapter. A product has the architecture that its requirement is allocated to physical components, e.g., cool down time of a computer is related to the fan, power supply, heat sink and so on. When a requirement is changed, the corresponding components may be triggered to change. It is not so simple to carry out ECs as most of them are generated by the interaction and coupling of components and the process is called change propagation.

Martin and Ishii [26] introduce the impacts of parameter changes on components to develop a design for various methods, and these impacts are evaluated with ratings. In this chapter, interaction between components is described through parameters. For example, component A_1 imposes geometry constraints or transfers forces, material or energy (which can be called parameters) to component A_2. It is able to learn about the interaction between components through the parameters. A model to analyze change propagation in terms of workload is outlined as shown in Fig. 5.1. First the relationships between change requirements and product components are mapped. Based on the relationships, the possible originating change components contributing to the change requirement can be acquired. Next the interactions between components in terms of parameters can be specified. For example, component A_4 is affected by parameters a_{11}, a_{12}, a_{13} of component A_1, and component A_6 is affected by parameters a_{21}, a_{22}, a_{23} of component A_2. Then the change propagation paths on the component level can be searched with an algorithm automatically, which is detailed in Sect. 5.4. After that, impacts of parameters on components are valued for the calculation of workload of different change propagation paths. Based on the workload, the change propagation analysis can be implemented.

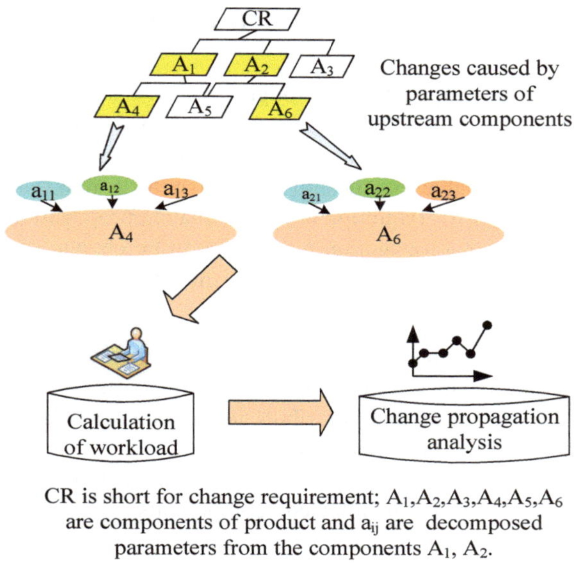

Fig. 5.1 Outline of workload-based change propagation analysis

An example is taken to show the process of change propagation analysis as shown in Fig. 5.2. A requirement is linked to several components and each of them can be an originating change component to satisfy the change requirement. An originating change component corresponds to different change propagation paths. Based on the workload, the optimal change propagation path with the least workload from an originating change component can be found. Then the optimal change propagation path for each originating change component can be acquired. Through comparing the workload of the optimal change propagation paths corresponding to the originating change components, the optimal change propagation path with the least workload can be obtained. After that, the corresponding originating change component can be selected as the optimal originating change component.

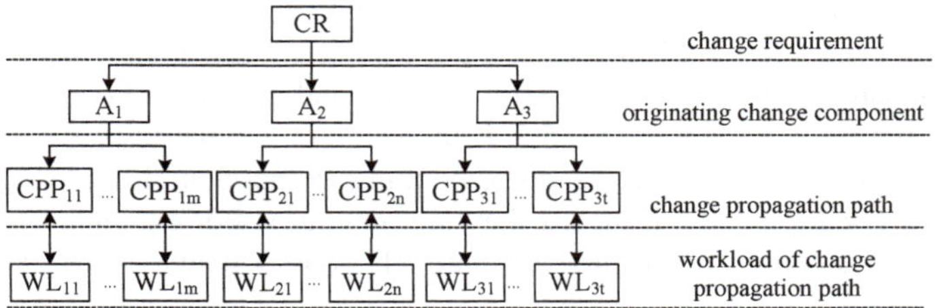

Fig. 5.2 Process of change propagation analysis

5.4 Searching Change Propagation Paths

In this chapter, change propagation is thought as a process during which initiating change components cause a series of subsequent changes. Initiating change component is chosen by designers to meet a change requirement and the change propagation aims to enable the design to evolve to a new stable status. The traceability achievement for requirement driven change propagation, which records and describes the change propagation process from the change requirement to the product components, can be depicted and interpreted by answering the following questions:

- What are the requirements of a product, i.e., low energy usage, reliability, low cost, etc., that are changed during the product lifecycle?
- What are the components that contribute to each change requirement?
- What parameters does each component contain?
- What are the component-component linkages constructed by parameters?
- What are the downstream affected components of an upstream change component?
- What are the logic relationships (AND, OR) between the downstream components according to the upstream change component?

5.4.1 Mapping from Change Requirements to Product Components

The Rational Unified Process defines a requirement as 'a condition or capability to which a system must confirm; either derived directly from user needs, or stated in a contract, standard, specification, or other formally imposed document' Diev [27]. Since components are medium to achieve product requirements, the corresponding components need to change inherently when a change requirement is demanded, e.g., the volume of reservoir needs to change when larger storage of a water cooler is demanded.

Information of the mapping from change requirements to product components can be gathered through staff interviews or by referring to change documents created by the company. With reference to Fig. 5.3, a product of three change requirements and composed of four components (modules) is assumed. The mapping relationships between change requirements and components are tabulated in the binary matrix as shown in Fig. 5.3. If the value in the matrix is "1", the change requirement from a row and the corresponding component from a column are related. If the value is "0", it means they are unrelated. For example, values from the first three columns of the first row are "1", components A_1, A_2 and A_3 are

Components

Change requirements	A_1	A_2	A_3	A_4
CR1	1	1	1	0
CR2	0	1	1	0
CR3	1	0	0	1

(a) Binary matrix for relationships between CRs and components

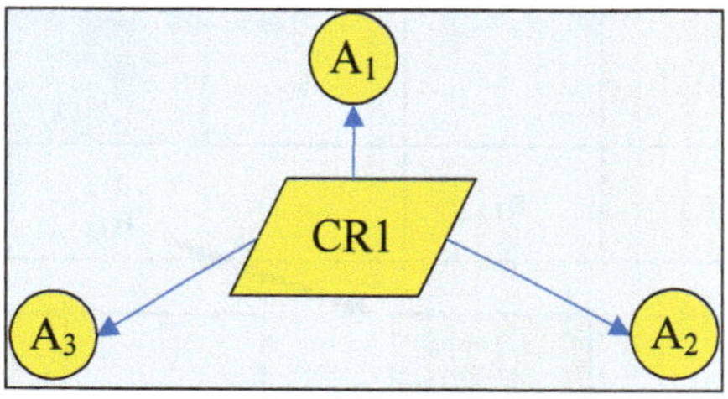

(b) Component changes caused by CR

Fig. 5.3 Change requirement-component matrix

responsible for change requirement 1. In other words, if change requirement 1 is demanded, one of components A_1, A_2 and A_3 needs to change. However, which one to change should be based on the assessment of impacts of change propagation caused by them.

5.4.2 Relationships Between Product Components

In this chapter, relationships between product components are captured through parameters, which are arranged in the component-component matrix. As shown in Fig. 5.4, relationships among the components A_1, A_2, A_3, and A_4 are taken as examples. Items in each column of the matrix are parameters belonging to a component and interaction between the components is indicated via parameters. For example, component A_3 affects component A_4 through its parameters a_{34}, a_{35} and component A_4 affects component A_3 through its parameters a_{41}, a_{42}. In this way, relationships between components can be acquired.

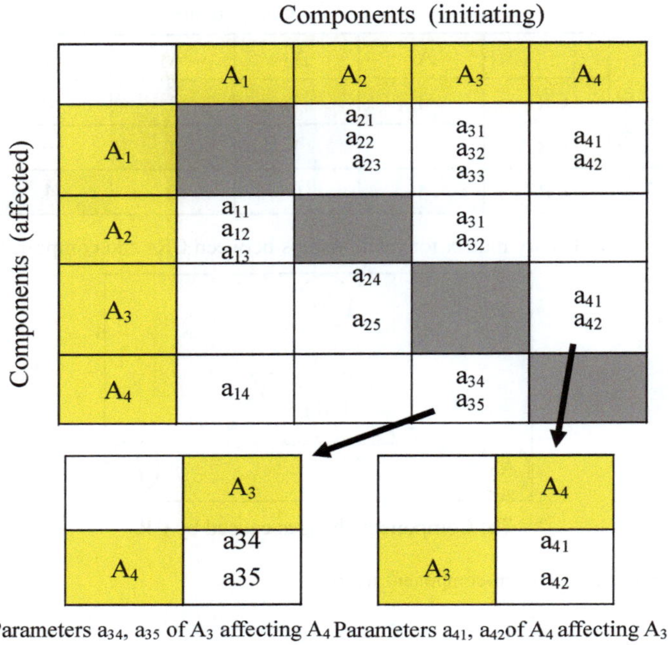

Fig. 5.4 Component-component matrix

5.4.3 Change Propagation Pattern

According to Fig. 5.5, in the relationship network of change components, a directed arrow is drawn from component A_1 to A_2, which means that the change of component A_1 can affect component A_2. When another arrow from component A_1 to A_3 exists, it means that the change of component A_1 may affect component A_3. This traditional type of network has been used for representing relationships between system components for further analysis of change propagation Yang and Duan [19] and how (why) changes propagating during engineering design Giffin et al. [18]. However, some logic relationships are not addressed during the change propagation

Fig. 5.5 Traditional change relationships

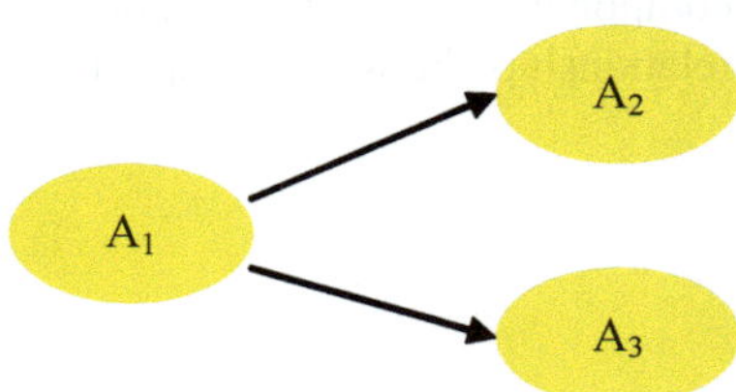

process, which cannot represent whether the downstream components are affected simultaneously or not. This is important for the workload determination of change propagation path. In this chapter, two types of logic change propagation pattern are applied.

Two types of junctions are used for searching change propagation paths: "OR" and "AND" relationships. "OR" represents the relationship that only one of the downstream components is affected by an upstream component at a time. To the contrary, "AND" represents the relationship that several downstream components are affected simultaneously by an upstream component. As shown in Fig. 5.6a, if the change of component A_1 causes components A_2 and A_3 to change simultaneously, the relationship between components A_2 and A_3 according to the upstream component A_1 is classified as "AND". Besides, the changes of components A_1, A_2, A_3 belong to the same change propagation path. In Fig. 5.6b, if component A_1 causes only one of the downstream components (A_2, A_3) to change, the relationship between components A_2 and A_3 corresponding to the upstream component A_1 is classified as "OR". Changes of components A_2 and A_3 belong to different change propagation paths.

A judgment is needed to determine the logic relationships between components. That is whether the affected components can be influenced by the same parameters of the upstream component. If the answer is yes, the logic relationship between the affected components is "AND". Otherwise the logic relationship is "OR". The parameters of a component affecting other components are represented in Eq. (5.1).

$$P\ (A_i,\ A_j) = \{a_{i1}, \ldots, a_{ik}, \ldots, a_{im}\} \tag{5.1}$$

where $P(A_i, A_j)$ is a set of parameters of component A_i which can affect component A_j. a_{ik} is a parameter in the set. m is the quantity of parameters of component A_i which can affect component A_j.

The relationships among components A_1, A_2, A_4 according to the upstream component A_3, as shown in Fig. 5.4, are taken as examples. According to Eq. (5.1), the sets of parameters for the components are concluded in Eqs. (5.2)–(5.4).

$$P\ (A_3,\ A_1) = \{a_{31}, a_{32}, a_{33}\} \tag{5.2}$$

$$P\ (A_3,\ A_2) = \{a_{31}, a_{32}\} \tag{5.3}$$

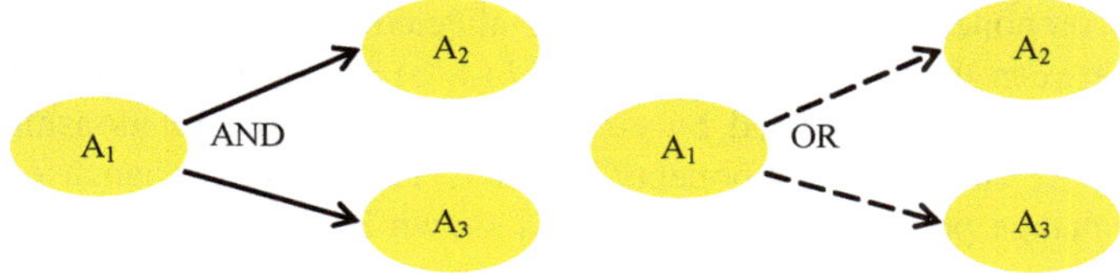

(a) AND relationship (b) OR relationship

Fig. 5.6 Logic relationships between components

$$P\ (A_3,\ A_4) = \{a_{34}, a_{35}\} \tag{5.4}$$

It is found that $P(A_3, A_1) \cap P(A_3, A_2) = \{a_{31}, a_{32}\} \neq \emptyset$, then the logic relationship between component A_1 and component A_2 corresponding to the upstream component A_3 is "AND". In this case, it is supposed that the change components A_1, A_2, A_3 are in the same change propagate path. It can also be found that $P(A_3, A_1) \cap P(A_3, A_4) = \emptyset$ and $P(A_3, A_2) \cap P(A_3, A_4) = \emptyset$, then the logic relationship between component A_4 and components A_1, A_2 is "OR". It is supposed that they are in different change propagation paths. Based on the conception, the possible change propagation paths can be searched which is detailed in the next section.

5.4.4 Searching Change Propagation Paths

Due to the complex relationships between components, it is difficult and effort-consuming to search change propagation paths manually. To address this problem, an algorithm is developed to search the change propagation paths more effectively.

In this chapter, components A_1, A_2, …, A_n are called 1, 2, …, n in short. In this way, each change requirement corresponds to a vector of sequential numbers and the relationship is represented in a form of Eq. (5.5).

$$(1\ \ 2\ \ \ldots\ \ i\ \ \ldots\ \ n)^T \rightarrow \mathrm{CR}j \tag{5.5}$$

where CRj is the jth change requirement. i means the ith component which contributes to the CRj. n is the quantity of components which contribute to the change requirement.

Besides, a logic relationship matrix, which is named as *LM*, is applied to express the logic relationships between components for the implementation of algorithm. The example shown in Fig. 5.4 is used to represent the logic relationship matrix as shown in Fig. 5.7. In the matrix, "0" represents that the components from the corresponding row and column are unrelated or the same. Other digits, in the matrix, represent the logic relationships between the affected components. If digits in a column are the same, the logic relationship between the affected components from the corresponding rows is "AND". Otherwise, it is "OR". For example, in Fig. 5.7, since *LM* (1, 3) = *LM* (2, 3) = 1, it means that, when component A_3 is changed, components A_1 and A_2 may be affected simultaneously and the relationship between them is "AND". Since *LM* (4, 3) = 2, it means another case that only component A_4 is affected by component A_3, and the relationship between component A_4 and other components is "OR". Thus, after a change propagation step, two change propagation paths exist as shown in Fig. 5.8.

In addition to encoding the product components and establishing the logic relationship matrix, a change propagation matrix named *CM*, which can change in

Components (affected) \ Components (initiating)	A_1	A_2	A_3	A_4
A_1	0	1	1	1
A_2	1	0	1	0
A_3	0	2	0	1
A_4	2	0	2	0

Fig. 5.7 Logic relationship matrix

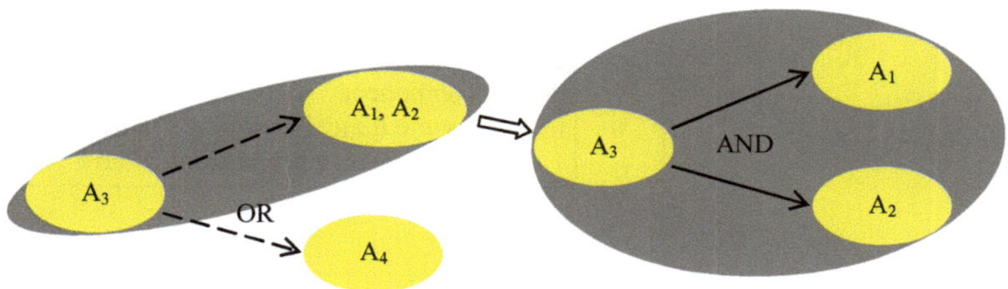

Fig. 5.8 Change propagation paths based on the logic relationship

size, is developed to represent the change propagation paths. In the change propagation matrix, each column vector represents a change propagation path. The direction from top to bottom is the direction of change propagation path and items in the direction are the affected components along the change propagation path. Thus the quantity of rows of the matrix is the number of change propagation steps and the quantity of columns of the matrix is the number of possible change propagation paths originating from an initiating change component.

Based on above, an algorithm is proposed to search change propagation paths automatically. The flowchart of searching change propagation paths is shown in Fig. 5.9. The process of the program execution is elaborated in the following steps:

Step 1: Select the *s*th originating change component as input of the algorithm

As shown in Fig. 5.3, if change requirement 1 is demanded, components A_1, A_2, A_3 need to change inherently to meet the requirement. Component A_3 is selected to illustrate the following steps.

Step 2: Acquire data from the logic relationship matrix

The *s*th (*k*th) column vector of the logic relationship matrix is acquired. For the component A_3, column vector $(1\ 1\ 0\ 2)^T$ is obtained from Fig. 5.7.

Fig. 5.9 Flowchart of searching propagation paths

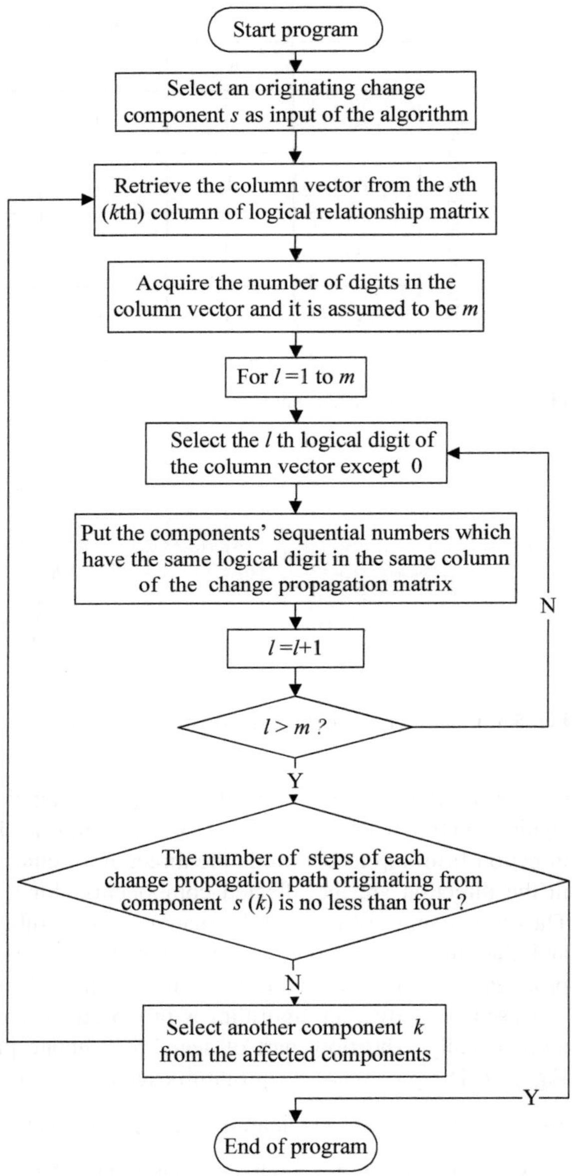

Step 3: Search the possible affected components in the next change propagation step

Digits in the *s*th (*k*th) column vector are traversed and 0 is excluded. Components with the same digit mean that they are affected by the upstream component simultaneously. Thus their sequential numbers are recorded in the same column and same row of the change propagation matrix. In terms of component A_3, after the first change propagation step, change propagation matrix *CM* (3) is shown in Eq. (5.6).

$$CM(3) = \begin{pmatrix} 3 & 3 \\ 1,2 & 4 \end{pmatrix} \tag{5.6}$$

where *CM* (3) indicates that it is the change propagation matrix for the originating change component A_3 affected by requirement 1. The first column vector $(3\ 1,\ 2)^T$ means that components A_1 and A_2 are affected simultaneously by component A_3 in the first change propagation step. The other column vector $(3\ 4)^T$ means another case that component A_4 is affected by component A_3. Thus two candidates of change propagation path exist for the originating change component A_3 after the first change propagation step. An electromotor can be taken as an example; two bearings are attached to an electromotor shaft to support its rotation and the shaft is protected in the housing. In one case, the diameter of the shaft and the inner diameters of the two bearings are the same for the function of rotation. Thus change of the shaft's diameter can cause both the bearings to change simultaneously, and the two bearings are supposed to be in the same change propagation path (AND relationship). In another case, the length of the shaft and the length of the housing are constrained. Thus, change of the shaft's length can cause the housing to change, and the housing is supposed to be in another change propagation path different from the bearings (OR relationship). In conclusion, the bearings and housing are affected by different parameters of the shaft, which leads to the different change propagation paths.

Step 4: Implement iteration process of change propagation

Note that in this work, the number of change propagation steps is four. This is to limit the reachability of change propagation to less than four steps, which is in line with the work by Giffin et al. [18] and Pasqual and de Weck [28] revealing that changes rarely propagate beyond four steps. According to this convention, if the quantity of change propagation steps in one change propagation path is less than four, select the *k*th component from the recently affected components as the initiating change component. Then go to *Step 2*. If the number of change propagation steps in every change propagation path is four, then go to *Step 5*.

Step 5: End of program

Based on the algorithm, the change propagation matrix from the originating change component A_3 is shown in Eq. (5.7). There is an iteration of change propagation. From the first column vector $(3\ 1,\ 2\ 2\ 1\ 2)^T$, it is found that component A_2 can affect A_1, and then component A_1 can affect A_2. The iteration is the repetition of change design due to different parameters affecting the components, which is a feature of change design processes that lends itself to modeling.

$$
alignedCM(3) = \left(\begin{array}{cccccccccccc}
3 & 3 & 3 & 3 & 3 & 3 & 3 & 3 & 3 & 3 & 3 & \\
1,2 & 1,2 & 1,2 & 1,2 & 1,2 & 1,2 & 1,2 & 1,2 & 1,2 & 1,2 & 1,2 & 1,2 \\
2 & 2 & 2 & 2 & 4 & 4 & 4 & 4 & 1 & 1 & 1 & 3 \\
1 & 1 & 3 & 3 & 1,3 & 1,3 & 1,3 & 1,3 & 2 & 2 & 4 & 1,2 \\
2 & 4 & 1,2 & 4 & 2 & 4 & 1,2 & 4 & 1 & 3 & 1,3 & 2
\end{array} \right.
$$
$$
\left. \begin{array}{cccccccccccc}
3 & 3 & 3 & 3 & 3 & 3 & 3 & 3 & 3 & 3 & 3 & 3 \\
1,2 & 1,2 & 1,2 & 1,2 & 4 & 4 & 4 & 4 & 4 & 4 & 4 & 4 \\
3 & 3 & 3 & 3 & 1,3 & 1,3 & 1,3 & 1,3 & 1,3 & 1,3 & 1,3 & 1,3 \\
1,2 & 1,2 & 1,2 & 4 & 2 & 2 & 4 & 1,2 & 1,2 & 1,2 & 1,2 & 4 \\
4 & 1 & 3 & 1,3 & 1 & 3 & 1,3 & 2 & 4 & 1 & 3 & 1,3
\end{array} \right) \tag{5.7}
$$

5.5 Workload-Based Analysis of Change Propagation

In this section, the concept of workload transformation is applied to calculate the workload of each change component, and then the total workload of each change propagation path is obtained. Based on the analysis results, the scheme of design change with least workload can be selected and implemented.

5.5.1 Workload Transformation Within a Change Propagation Step

In this chapter, the workload that has to be done by each coupled task is assumed as a linear combination of the workload done on other tasks in the previous design change stage and the workload transformed between components is assumed to be diminishing in line with that changes are absorbed progressively during the change propagation process. Besides, five assumptions for the linear algebraic analysis of design change are made as follows:

(1) Task is defined as the activity of changing the parameters of product components.
(2) The design change workload of interdependent product components is transformed through parameters.
(3) All tasks of design change are executed at every stage (i.e., completely parallel process).
(4) The quantity of design change work of component decreases progressively with the increase of change propagation step.
(5) The workload transformation values (i.e., the strength of dependency) do not vary with time.

Based on the above assumptions, the parameters' impacts of one component on other components are valued and the values are formatted in the component-component matrix. The example in Fig. 5.4 is used to represent the general form of the component-component matrix as shown in Fig. 5.10. Workload transformation between components is computed with a formula as shown in Eq. (5.8).

$$W_m(A_j) = \left[1 - \prod_{k=1}^{n(A_i,A_j)} \left(1 - \frac{1}{N(A_i)} i_{a_{ik},A_j}\right)\right] W_{m-1}(A_i) \tag{5.8}$$

where $n(A_i, A_j)$ parameters are regarded as medium of linkages between components A_i and A_j. From the perspective of parameters, if A_i is changed, at least, one of the $n(A_i, A_j)$ parameters will cause A_j to change. Besides, $N(A_i)$ is the total number of parameters from component A_i that may affect other components. As the impact between components is measured as a probabilistic workload, $1/N(A_i)$ acts as a possibility coefficient and a weight of parameter's impact on other components. Impact of the kth parameter a_{ik} on component A_j is valued as ia_{ik}, A_j. $W_{m-1}(A_i)$, $W_m(A_j)$ are the workload to implement changes on components A_i and A_j respectively in the $(m-1)$th and mth change propagation step.

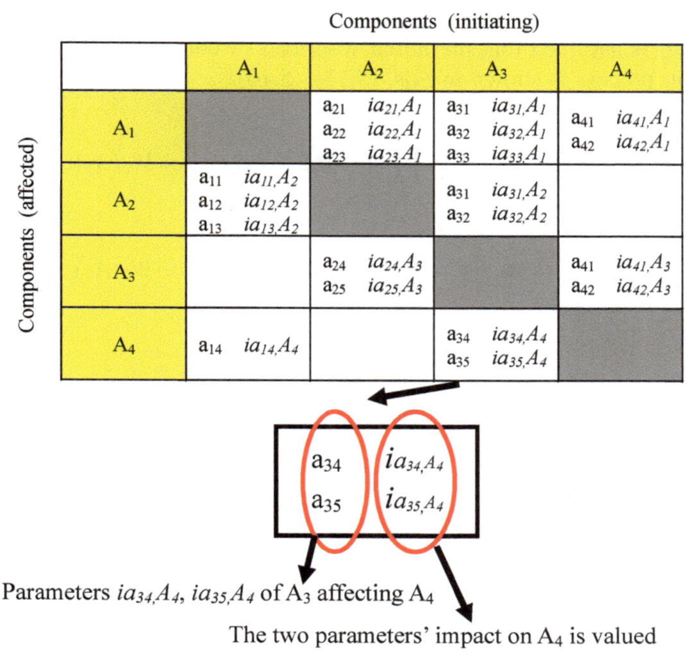

Fig. 5.10 Component-component matrix with parameters' impacts valued

For instance, the amount of rework of initiating change component A_3 is set as $W_0(A_3)$, then the amount of rework of A_4 caused by A_3 is expressed in Eq. (5.9).

$$W_1(A_4) = \left[1 - (1 - 0.2 \times i_{a_{34},A_4}) \times (1 - 0.2 \times i_{a_{35},A_4})\right] \times W_0(A_3) \tag{5.9}$$

5.5.2 *Workload for a Change Propagation Path*

According to the change propagation paths and workload transformation formula between components, the workload of each change propagation path is computed by summing the workload of change components in the change propagation path and it can be expressed in Eq. (5.10).

$$S(P_h) = \sum_{A_i \in P_h} W_t(A_i) \tag{5.10}$$

where $S(P_h)$ is the workload of the hth change propagation path P_h. The path P_h consists of some components, including component A_i. $W_t(A_i)$ is the workload of component A_i in the tth change propagation step.

For example, the change propagation path represented by the column vector (3 1,2 2 1 2)^{T} in the change propagation matrix, as shown in Eq. (5.7), is taken as an example. It is assumed that the initial workload of component A_3 is $W_0(A_3)$. The calculation process is shown in Eqs. (5.11)–(5.16).

$$W_1(A_1) = \left[1 - \prod_{i=1}^{3} (1 - 0.2 \times i_{a_{3i},A_1})\right] \times W_0(A_3) \tag{5.11}$$

$$W_1(A_2) = \left[1 - \prod_{i=1}^{2} (1 - 0.2 \times i_{a_{3i},A_2})\right] \times W_0(A_3) \tag{5.12}$$

$$W_2(A_2) = \left[1 - \prod_{i=1}^{3} (1 - 0.25 \times i_{a_{1i},A_2})\right] \times W_1(A_1) \tag{5.13}$$

$$W_3(A_1) = \left[1 - \prod_{i=1}^{3} (1 - 0.2 \times i_{a_{2i},A_1})\right] \times W_2(A_2) \tag{5.14}$$

$$W_4(A_2) = \left[1 - \prod_{i=1}^{3} (1 - 0.25 \times i_{a_{1i},A_2})\right] \times W_3(A_1) \tag{5.15}$$

$$S(P_1) = W_0(A_3) + W_1(A_1) + W_1(A_2) + W_2(A_2) + W_3(A_1) + W_4(A_2) \tag{5.16}$$

5.5.3 Change Propagation Analysis and Recommendation

As described in Sect. 5.4, more than one change propagation paths as candidates may originate from an initiating change component. If the workload of a path is less than others', it is favorable; otherwise, it is unfavorable. Thus the change propagation path with the least workload shall be selected.

An example is given to illustrate the method. As shown in Fig. 5.3, change requirement 1 is demanded. Then the corresponding components are changed to meet the requirement. Components A_1, A_2, A_3 are assumed to be the initiating change components. As workload is related with the consumption of resources and development time, the workload is expressed as the product of the quantity of involved designers and the development time. The skills of the designers are assumed to be similar. For example, if a design change is done by 3 designers with 10 h, the workload is 30 ($3 \times 10 = 30$). In this example, these components' initiating change workload is assumed as 10. Component-component matrix in Fig. 5.10 is transformed into the valued component-component matrix as shown in Fig. 5.11. Parameter's impacts on other components are valued as 0.25, 0.50 or 0.75, which are estimated according to low, medium or high impact on the affected components. The degree of impact can be described as follows: if a small change in the parameter requires a change in the component, then the parameter has a high impact on the component; if a large change in the parameter requires a change in the component, then the parameter has a low impact on the component; and the medium impact ranges between the low and high impact. Component A_3 is used to show the process of computation of workload and change propagation analysis. As shown in Eq. (5.7), the change propagation paths of initiating change component A_3 are represented. With the computation method described above, workload of each change propagation path is calculated. The results are shown in Fig. 5.12.

Components (initiating)

Components (affected)	A_1	A_2	A_3	A_4
A_1		a_{21} 0.25 a_{22} 0.50 a_{23} 0.25	a_{31} 0.25 a_{32} 0.25 a_{33} 0.25	a_{41} 0.25 a_{42} 0.50
A_2	a_{11} 0.75 a_{12} 0.50 a_{13} 0.25		a_{31} 0.25 a_{32} 0.50	
A_3		a_{24} 0.25 a_{25} 0.25		a_{41} 0.50 a_{42} 0.50
A_4	a_{14} 0.50		a_{34} 0.75 a_{35} 0.50	

Fig. 5.11 Impact-valued component-component matrix

Path1	Workload	…	…	Path11	Workload	…	…	Path18	Workload	…	…	Path24	Workload
A_3	10			A_3	10			A_3	10			A_3	10
↓				↓				↓				↓	
A_1,A_2	3.778			A_1,A_2	3.778			A_4	1.9			A_4	1.9
↓				↓				↓				↓	
A_2	0.3306	…	…	A_1	0.5918	…	…	A_1,A_3	1.4844	…	…	A_1,A_3	1.4844
↓				↓				↓				↓	
A_1	0.103			A_4	0.111			A_2	0.115			A_4	0.1579
↓				↓				↓				↓	
A_2	0.0181			A_1,A_3	0.0867			A_3	0.0218			A_1,A_3	0.1234
Total workload	14.2297	…	…	Total workload	14.5675	…	…	Total workload	13.5212	…	…	Total workload	13.6657

Fig. 5.12 Workload of different change propagation paths from component A_3

The total workload of each change propagation path, as shown in Fig. 5.12, is plotted as a column graph, which is shown in Fig. 5.13 to analyze the change propagation. The total workload of the 18th change propagation path is the least. Thus the 18th change propagation path is the optimal change propagation path originating from component A_3. The total workload of the 11th change propagation path is the most. Thus this solution of design change should be avoided as far as possible. Correspondingly, change propagation originating from A_1, A_2 can be analyzed similarly as that of A_3. Based on above, recommendation is that the change propagation path of the least workload should be given priority and efforts of implementing changes should be put on the components in the change propagation path. Besides, it is found that the workload of the optimal change propagation paths originating from components A_1, A_2, A_3 is about 12.23, 12.56, and

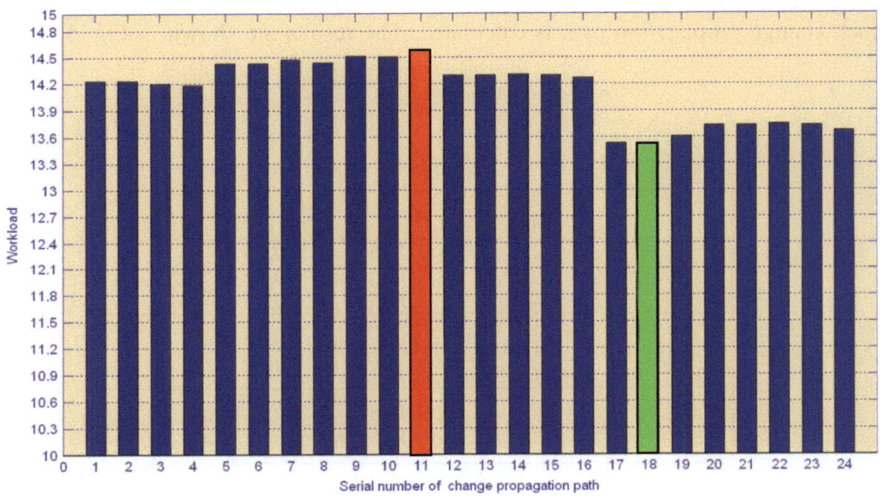

Fig. 5.13 Workload-based change propagation analysis from component A_3

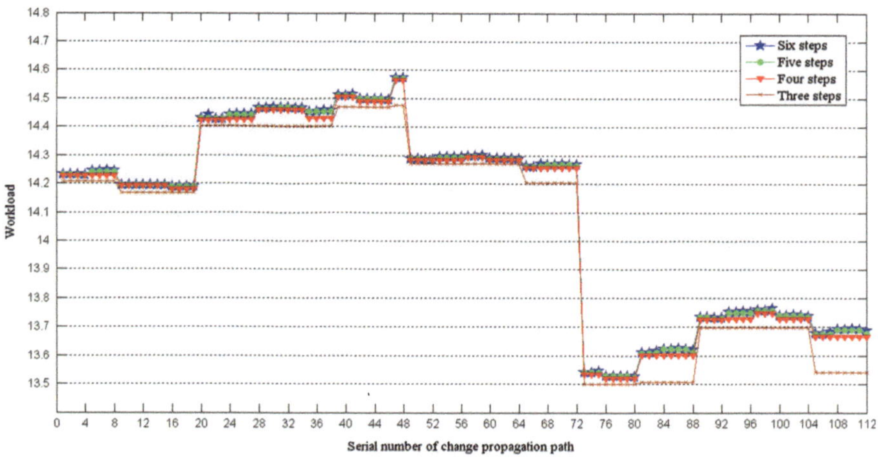

Fig. 5.14 Workload distribution of change propagation paths with different steps

13.52 respectively. According to the workload of optimal change propagation paths, A_1 is identified as the optimal originating change component.

It can be seen from Fig. 5.12 that four change propagation steps are assumed. A comparison is made to explain that the assumption is appropriate. Component A_3 is used to demonstrate the comparison process. As show in Fig. 5.14, curves describing the workload distribution of change propagation paths (with different steps) originating from component A_3 are demonstrated. It can be seen that the curves of more than four steps (e.g., the curves of six/five steps) are almost the same as the one of four steps. Besides, the curve of less than four steps (e.g., the curve of three steps) is similar to the one of four steps. Thus four propagation steps can be regarded as a representative for the simulation of change propagation process.

5.6 Application of the Method for Product Design

The proposed method was applied to a blowing and suction machine from an electronics company, which is shown in Fig. 5.15, to examine how the method can be used to analyze the change propagation during the process of design change. The product was simplified and main parts (modules) were constructed to demonstrate the initial evaluation of the method. In the following subsections, the modeling process, the process of searching change propagation paths and the process of analyzing change propagation are described.

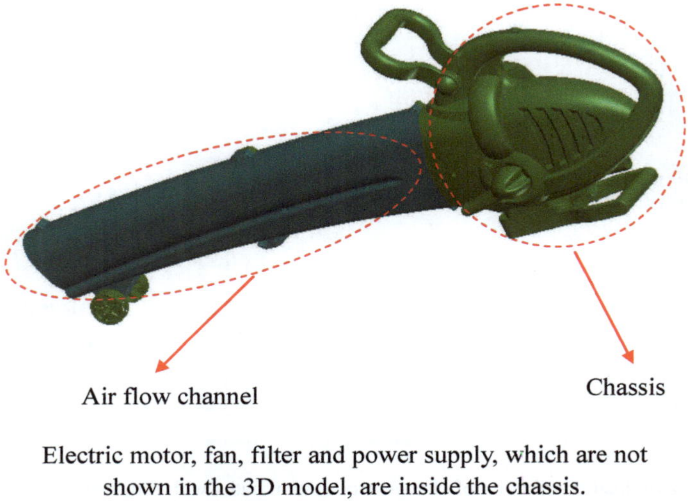

Fig. 5.15 Blowing and suction machine

5.6.1 Modeling Process

In the modeling process, all the data of the blowing and suction machine was acquired through consulting the designers from the electronics company, who had worked on products similar to the machine, and a standard process was made to improve the modeling efficiency as far as possible. First, the blowing and suction machine was divided into six product components (modules) with seven possible change requirements, and they are shown in Tables 5.1 and 5.2. Second, the first kind of relationship, expressed in the change requirement-component matrix, was elicited as shown in Fig. 5.16. Third, each designer was asked to provide the parameters of a component (module) on which they had worked, and then the linkages between components were acquired through analyzing the parameters of every two components. Finally, all these linkages were arranged in the valued component-component matrix as shown in Fig. 5.17. Based on the valued component-component matrix and the conception described in Sect. 5.4.3, the logic relationships between components were obtained as shown in Fig. 5.18.

Table 5.1 Components of the machine

No.	Component name	Abbrev.
1	Electric motor	EM
2	Fan	FA
3	Chassis	C
4	Filter	FI
5	Air flow channel	AFC
6	Power supply	PS

Table 5.2 Change requirements of the machine

No.	Change requirement	Abbrev.
1	High flow rate	HFR
2	Compact	C
3	Low vibration	LV
4	Low noise	LN
5	Rapid cooling	RC
6	Low energy usage	LEU
7	Low cost	LC

Components

Change requirements	EM	FA	C	FI	AFC	PS
HFR	1	1	0	0	1	1
C	0	0	1	0	1	0
LV	0	1	0	0	1	0
LN	0	1	0	1	1	0
RC	0	1	1	0	1	0
LEU	1	1	0	0	0	1
LC	1	0	1	0	1	1

Fig. 5.16 Change requirement–component matrix of the machine

5.6.2 *Searching Change Propagation Paths*

The high flow rate, as an important performance metric, was selected as the change requirement to demonstrate the process of searching change propagation paths. According to the data in Fig. 5.16, one of the electric motor, fan, air flow channel and power supply should be selected to meet the change requirement (high flow rate). According to the algorithm of searching change propagation paths described in Sect. 5.4, change propagation matrices for the originating change components (electric motor, fan, and air flow channel and power supply) can be generated.

5.6.3 *Workload-Based Analysis of Change Propagation*

According to the method of assigning initiating change workload described in Sect. 5.5.3, the workload of the four originating change components can be acquired and the results are 10, 9, 11, and 15 respectively. Workload of change

Components (initiating)

Components (affected)	EM	FA	C	FI	AFC	PS
EM		pressure curve 0.50 z dim 0.50				voltage 0.75 p-p voltage 0.50
FA	x dim 0.50 z dim 0.50			opening rate 0.50 z loc 0.25	x dim 0.50 y dim 0.50 z dim 0.50	
C	weight 0.25 mounting holes 0.25 x dim 0.50 y dim 0.50 z dim 0.50 x loc 0.25 y loc 0.25 z loc 0.50	x dim 0.50 y dim 0.50		x dim 0.25 y dim 0.25	x dim 0.75 y dim 0.75 x loc 0.25 y loc 0.25	
FI		pressure curve 0.50 x loc 0.25 y loc 0.25			x dim 0.25 y dim 0.25 z dim 0.25	
AFC		flow rate 0.50 airflow direc 0.50 z loc 0.50		x dim 0.25 y dim 0.25		
PS	Rated power 0.75	flow rate 0.50		opening rate 0.50		

Fig. 5.17 Impact-valued component-component matrix of the machine

propagation paths originating from electric motor is shown in Fig. 5.19 and the corresponding change propagation analysis is shown in Fig. 5.20. Figure 5.19 shows the total workload of different change propagation paths as well as the workload of each affected component from electric motor. It can be seen from Fig. 5.20 that the 26th change propagation path has the least workload, suggesting that this change propagation path is the optimal change propagation path. Besides, the change propagation originating from fan, air flow channel and power supply can be analyzed similarly as that of electric motor.

The recommendation for the selection of change propagation path is that giving priority to the one with the least workload and selecting the components in the change propagation path to implement changes. Besides, the result from the change propagation analysis shows that the workload of the optimal change propagation paths originating from electric motor, fan, air flow channel and power supply is

Components	EM	FA	C	FI	AFC	PS
EM	0	1	0	0	0	1
FA	1	0	0	1	1	0
C	1	2	0	2	1	0
FI	0	1	0	0	1	0
AFC	0	3	0	2	0	0
PS	2	3	0	1	0	0

Fig. 5.18 Logic relationship matrix of the machine

Path1	Workload	…	…	Path13	Workload	…	…	Path20	Workload	…	…	Path26	Workload
EM	10			EM	10			EM	10			EM	10
↓				↓				↓				↓	
FA,C	3.9719			FA,C	3.9719			FA,C	3.9719			PS	0.8333
↓				↓				↓				↓	
EM,FI	0.2326	…	…	AFC,PS	0.2302	…	…	AFC,PS	0.2302	…	…	EM	0.4427
↓				↓				↓				↓	
FA,C	0.0464			FA,C,FI	0.1296			EM	0.0319			PS	0.0369
↓				↓				↓				↓	
EM,FI	0.0027			EM,FI	0.0099			PS	0.0027			EM	0.0196
Total workload	14.2536	…	…	Total workload	14.3416	…	…	Total workload	14.2367	…	…	Total workload	11.3325

Fig. 5.19 Workload of different change propagation paths from electric motor

about 11.33, 11.08, 17.62 and 16.01 respectively. Based on the results, fan is the optimal originating change component to satisfy the high flow rate as it causes the least workload of design change.

5.7 Initial Evaluation

In an effort to test out the usefulness of such analysis, the work was presented to two experienced designers from the electronics company, who had worked on the blowing and suction machine. They were invited to participate in the initial evaluation of the method. The product model of the blowing and suction machine established in Sect. 5.6 was provided to the designers. As they had rich experiences in electronics design, their comments for the improvement of the machine's model

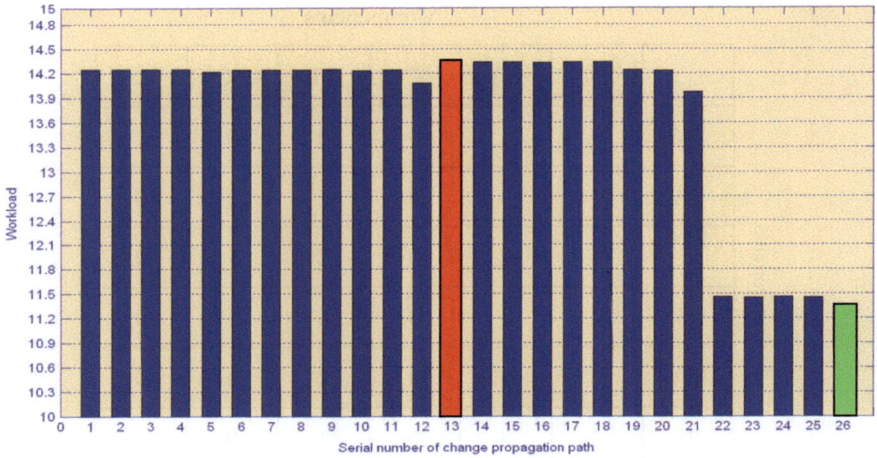

Fig. 5.20 Workload-based change propagation analysis from electric motor

were given more weight. Then the analysis process of the method was studied by them independently and implemented on the blowing and suction machine. It was found that the method could help to better focus discussions on strategies of design change. One designer pointed out that ‘As the fan of the blowing and suction machine causes the least workload to satisfy the high flow rate, I will first consider it as the originating change component.’ In addition, the other designer also commented that ‘This method gives you a static analysis of a given product. This is good and useful to some extent. A database, storing the data of schemes of design change, can be established, which can further facilitate the knowledge-based design.’ However, the first designer also questioned that ‘Product design is an evolving process, thus the relationships within the product are changing too. Then should the digits in the relationship matrices still be constant?’

5.8 Discussion

The presented method was tested against the case of blowing and suction machine, and the method was also initially evaluated by two experienced designers. In this section, further discussions on the proposed method are provided.

There is a scenario that components are changed to satisfy a requirement, and meanwhile leads to changes in other requirements. Two kinds of circumstances exist: one is that changes of components positively affect other requirements (i.e., better satisfy other requirements); the other is that changes of components negatively affect other requirements (dissatisfy other requirements). The first kind of circumstance is welcomed for the enhancement of a product. To address the second kind of circumstance, a solution is that selecting some other initiating components to compensate for the loss of affected requirements and implement combined analysis of change propagation with the developed method. Besides, if the affected requirements are vital, then the solution of design change should be modified and the change propagation path with the second least workload should be selected for the design change. This process can be implemented continuously until the optimal scheme is found.

In this chapter, both the consumption of resources and development time are considered for the definition of workload. Chu et al. [29] measure the workload of task in terms of its development time. However, the differences of development time between related tasks may lead to the completion of tasks at different times. For example, the execution of a downstream task cannot be implemented until the upstream task is completed, which can cause a waste of resources. To analyze the impact of design change comprehensively, both resources and development time are considered for the development of schemes of design change in this chapter. From the perspective of workload, more workload implies more consumption of resources and development time. Less workload can save the resources and shorten the cycle of product, which will improve the product competition in the market.

5.9 Conclusion and Future Work

This chapter presents a theoretical method that can be used to analyze change propagation in terms of workload and an initial evaluation of the method is carried out. According to the workload of different change propagation paths, the change propagation can be analyzed and the optimal scheme of design change can be selected and implemented efficiently. The method considers the change requirement as the cause of change propagation and the changes of components as the solutions to satisfy the change requirement. Change propagation paths from an originating change component are assessed in terms of total workload. Two types of relationship matrix are applied in this chapter: requirement-component matrix and component-component matrix. The possible originating change components to meet a change requirement are selected through checking the change requirement-component matrix. The logic relationships between components are applied to search the change propagation paths with a developed algorithm. The workload of different change propagation paths is calculated according to the valued component-component matrix and possible change propagation paths. Then the optimal change propagation path from an originating change component can be acquired through comparing the total workload of different change propagation paths. Through comparing the workload of optimal change propagation paths corresponding to the originating change components, the optimal originating change component can be obtained. By referring to this method, a blowing and suction machine is applied to test the method initially.

Some assumptions are contained in the theoretical method, the validity of which needs to be further explored and more work remains to be done. At the same time, some areas have been identified for future development.

(1) The case studied in this chapter is relatively simple with only a few coupled components (modules), while the number of parameters, linkages between components, and the complexity of linkages increase remarkably in complex designs. Thus, the extension of this method to complex designs may bring new research opportunities and more tests against products will be carried out to evaluate this method.
(2) The impacts between components are based on experts' opinions or from the documents (records of estimations given by designers working on similar engineering systems). Although this approach allows one to capture implicit knowledge from the engineering designers, the data collected can be subjective and the elicitation process can be time consuming when the product has a number of components. Thus, more efforts are still required to improve the elicitation of reliable change data efficiently.
(3) Generally, the elements in the real design disperse in different computing models. The component-component matrix is planned as an integration framework to organize the component linkages stored in different computing models, so that the change propagation can be analyzed in a comprehensive context.

References

1. Tang DB, Xu RH, Tang JC et al (2008) Design structure matrix-based engineering change management for product development. Int J Internet Manuf Serv 1(3):231–245
2. Tang DB, Xu RH, Tang JC et al (2010) Analysis of engineering change impacts based on design structure matrix. J Mech Eng 46(1):154–161 (in Chinese)
3. Li YL, Zhao W, Shao XY (2012) A process simulation based method for scheduling product design change propagation. Adv Eng Inform 26(3):529–538
4. Morkos B, Shankar P, Summers JD (2012) Predicting requirement change propagation, using higher order design structure matrices: an industry case study. J Eng Des 23(12):905–926
5. Ulrich KT, Eppinger SD (1995) Product design and development, 1st edn. McGraw-Hill, New York
6. Pahl G, Beitz W (1998) Engineering design: a systematic approach, 2nd edn. Springer, London
7. Otto K, Wood K (2000) Product design. Prentice Hall, Upper Saddle River
8. Ullman DG (2003) The mechanical design process, 3rd edn. McGraw-Hill, New York
9. Hull E, Jackson K, Dick J (2005) Requirements engineering, 2nd edn. Springer, London
10. Ahmed S, Yethish K (2007) Engineering change during a product's lifecycle. In: International conference on engineering design, ICED'07, Paris, France
11. Fricke E, Gebhard B, Negele H et al (2000) Coping with changes: causes, findings, and strategies. Syst Eng 3(4):169–179
12. Maier AM, Langer S (2011) Engineering change management report 2011: survey results on causes and effects, current practice, problems, and strategies in Denmark. DTU Management

Engineering, Department of Management Engineering, Technique University of Denmark, Copenhagen, Denmark
13. Bloor MS, Owen J (1991) CAD/CAM product data exchange: the next step. Comput Aided Des 23(4):237–243
14. Gilbert M (1991) Application protocols. In: Seminar proceedings of the product data exchange for the 1990s. US National Institute of Standards and Technology, Gaithersburg
15. Wang Y (1991) Three schema implementation in STEP: standard for the exchange of product model data. In: Computers and engineering. ASME, New York, pp. 1–5
16. Cohen T, Navathe SB, Fulton RE (2000) C-FAR, change favorable representation. Comput Aided Des 32(5/6):321–338
17. Jan C, Till D, Christian K, Soren W et al (2007) Change impact and risk analysis (CIRA)—combining the CPM/PDD theory and FMEA-methodology for an improved engineering change management. In: International conference on engineering design, ICED'07, Paris, France
18. Giffin M, de Weck O, Bounova G et al (2009) Change propagation analysis in complex technical systems. J Mech Des 131(8):081001
19. Yang F, Duan GJ (2012) Developing a parameter linkage-based method for searching change propagation paths. Res Eng Des 23(4):353–372
20. Clarkson PJ, Simons C, Eckert C (2004) Predicting change propagation in complex design. J Mech Des 126(5):788–797
21. Edwin CYK, Nicholas HMC, Clarkson PJ (2013) A technique to assess the changeability of complex engineering systems. J Eng Des 24(7):477–498
22. Oh S, Park B, Park S et al (2007) Design of change-absorbing system architecture for the design of robust products and services. Hum Comput Interact 4553:1110–1119
23. Li YL, Zhao W (2014) An integrated change propagation scheduling approach for product design. Concurr Eng Res Appl 22(4):347–360
24. Pahl G, Beitz W, Feldhusen J et al (2007) Engineering design: a systematic approach, 3rd edn. Springer, London
25. Smith RP, Eppinger SD (1997) Identifying controlling features of engineering design iteration. Manag Sci 43(3):276–293
26. Martin MV, Ishii K (2002) Design for variety: developing standardized and modularized product platform architectures. Res Eng Des 13(4):213–235
27. Diev S (2007) Requirements development as a modeling activity. ACM SIGSOFT Softw Eng Notes 32(2):1–3
28. Pasqual MC, de Weck OL (2012) Multilayer network model for analysis and management of change propagation. Res Eng Des 23(4):305–328
29. Chu CC, Chen SS, Zheng PE (2006) Project scheduling model based on dependency structure matrix. Comput Integr Manuf Syst 12(10):1591–1595 (in Chinese)

Chapter 6
Matrix-Based Engineering Change Management of Product Design Using MBD Technique

6.1 Introduction and Synopsis

Engineering change management is a significant activity in industry and the importance of change management is appraised. Different reports suggest that ECs use around one-third of the engineering design capacity [1–3]. Since the later a change or the impact of a change is detected, the more expensive it becomes to undertake. Resource-intensive/costly engineering changes and their subsequent change propagation should be avoided earlier. Besides, unmanaged response to changes can cause the product to drift from its optimal design state and to demonstrate progressively poor performance [4, 5]. Most probable way to prevent this anomaly is to propose methods and tools which can detect changes efficiently and analyze the impact of change propagation proactively.

Traditionally, engineers detect changes and predict change propagation based on abundant product data and complex product relations scattered in different files, which is time-consuming and error-prone. Meanwhile, the product development time is continuously decreasing, which further increases the burden on the engineers. As described in Chap. 5, there are many researches about engineering change management. However, few methods combine the product change-related data with the CAD models, which can facilitate the design change process automation and improve the efficiency of the change management. Jorge et al. [6] propose an annotation-centered design method to enhance the product model reuse. However, it is mainly developed to represent component functions, and a small amount of product data is recorded in the CAD model. MBD (model-based definition) is a new way of managing engineering and business processes using CAD model as the complete source of information for design, production, distribution, technical documentation, services, and the entire product lifecycle [7], and can contribute to the efficient implementation of design change. In this chapter, product change-related data is recorded in the introduced DCMBD (design change-oriented model-based definition) model with a predefined data structure. Besides, with the

D. Tang et al., *Matrix-Based Product Design and Change Management*,
DOI 10.1007/978-981-10-5077-0_6

product MBD models, the relations and initial changes of product can be captured automatically, rather than acquired manually by designers with experiences.

The remainder of the chapter is organized as follows. In Sect. 6.2, the definition of DCMBD model is described. In Sects. 6.3 and 6.4, algorithms to acquire the two types of product change (parameter change and topology face change) are presented, respectively. Section 6.5 introduces the relation models for parameter and topology face, and Sect. 6.6 presents the change propagation patterns and design change process. In Sect. 6.7, a prototype system for product design change is developed and an application is demonstrated to show how the MBD-based method can be applied to the product design change. Subsequently, discussions about the method and its extension are given. Finally, conclusions of this work are provided.

6.2 Design Change-Oriented MBD Model

MBD model has the advantage of collecting and managing product data inside three-dimensional CAD model. And some MBD methods have already been applied in product design. For example, Alemanni et al. [7] introduce three types of MBD for different product management scenarios. However, most of the MBD methods are aimed at improving the communication of design intent and rationale between designers. In this chapter, to acquire the product design changes automatically, the product design data is required to be embedded in the MBD models. Besides, product relations are also acquired automatically. Thus the relations between parameters and constraints between topology faces should also be defined in the interfaces of MBD models. To meet these requirements, the data structure of MBD model is studied in this section. Referring to Alemanni et al. [7], the data structure of part MBD model and product MBD model is standardized, and the details are listed as shown in Fig. 6.1. The description in the green boxes records the information of the part MBD model and the description in the red boxes records the information of the product MBD model. This type of MBD model focuses primarily on the design change-related data representation, and is named as DCMBD model in this chapter. With the DCMBD model, the change-related data of the product is stored in the three-dimensional CAD model. Since modern CAD software is able to manage the data of DCMBD model as geometrical entity, it can contribute to the automation and standardization of design change management.

Besides, the representation of DCMBD model for part and product can be concluded as shown in Eqs. (6.1) and (6.2). "ID" is used to identify the product/part; "CG" records the geometrical features and topological elements of part, which can be used to acquire the part structural changes; "M&T" records the part material-related information for material change acquisition; "FT&A" mainly records the annotations and dimensions of part; "R" and "GC" record the relations of the two types of product item for change propagation analysis; "LMD" records the version of change database, notes of product design change, and technological data changes etc. In this chapter, parameter is defined as engineering attribute and

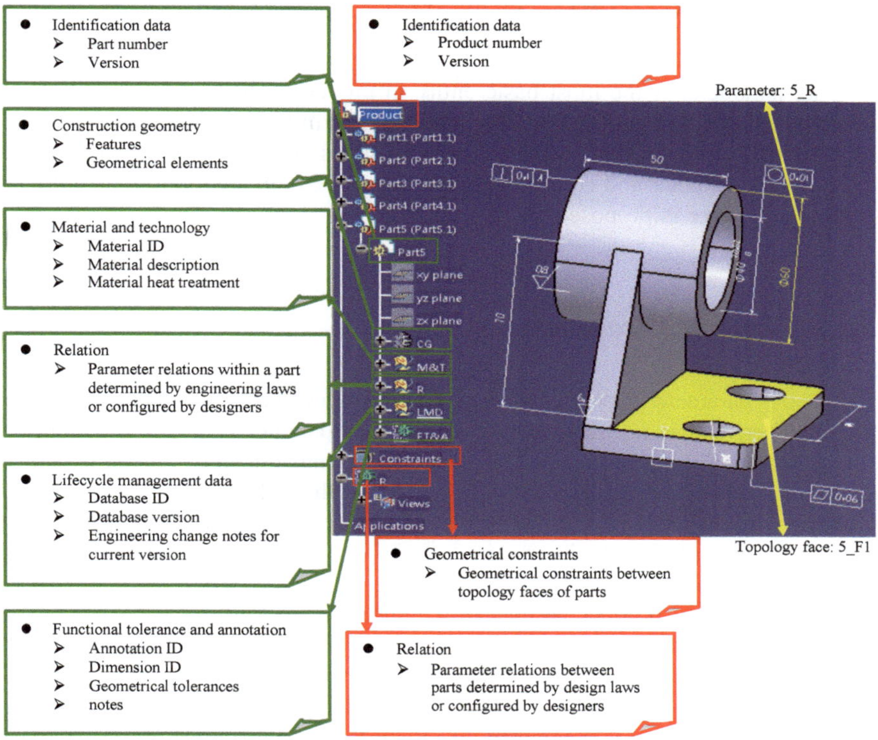

Fig. 6.1 Example of design change-oriented MBD model

can be described with variable in the product design. Topology face is defined as geometric face bounding a part. For example, diameter "⌀60" of the Part 5 in Fig. 6.1 can be regarded as a parameter (*5_R*) and the yellow face can be regarded as a topology face (*5_F1*). If the value of the diameter "60" is changed, the parameter *5_R* is changed. And if the shape, dimension, etc. of the yellow face are changed, the topology face *5_F1* is changed. The parameter change and topology face change will be further demonstrated in Sects. 6.3 and 6.4, respectively.

$$\mathrm{M}(\text{part}\ i) = \{\mathrm{ID} \oplus \mathrm{CG} \oplus (\mathrm{M\&T}) \oplus (\mathrm{FT\&A}) \oplus \mathrm{R} \oplus \mathrm{LMD}\} \tag{6.1}$$

$$\mathrm{M}(\text{product}\ j) = \{\mathrm{ID} \oplus \mathrm{M}(\text{part 1}) \oplus \cdots \oplus \mathrm{M}(\text{part}\ n) \oplus \mathrm{R} \oplus \mathrm{GC}\} \tag{6.2}$$

where M(part *i*) and M(product *j*) represent the part DCMBD model and product DCMBD model, respectively; ID: identification data; CG: construction geometry; M&T: material and technology; FT&A: functional tolerance and annotation; R: relation; GC: geometrical constraint; LMD: lifecycle management data.

6.3 Change Acquisition in Terms of Product Parameter

Product parameter is the most basic entity in product design, and the daily engineering decision making deals mainly with the determination of engineering values of parameters [8]. In this section, first, two types of parameter are introduced: exogenous parameter and constrained parameter. Next, an algorithm to acquire the parameter change is proposed.

6.3.1 Type of Parameter

Exogenous parameter is defined as the parameter which can be modified directly by the designer, and is not dependent on any other parameters, e.g. the "diameter" (*d*) or "material" (gray iron: HT 250) of a piston. Constrained parameter is defined as the parameter which can only be satisfied through exogenous parameters according to design laws, e.g. the stroke of a slider. Parameter is named with an "ID" and is recorded in the interfaces ("M&T" and "FT&A") of the part DCMBD model with a standard data format.

In this chapter, product performance indices are regarded as the sources of requirement change. The product performance indices can be described mainly through the constrained parameter and the constrained parameter can be modified through exogenous parameters. Thus, design change of parameter is aimed at modifying the exogenous parameters to satisfy the changed constrained parameter according to the requirement change. For example, the slider stroke of a press machine can be classified as a constrained parameter and is an important performance index. Equation (6.3) is the corresponding formulation of design law in the press machine design. It can be seen that the exogenous parameters R and L may be changed to satisfy the changed slider stroke S.

$$S = R \cdot \left[(1 - \cos \alpha) + \frac{R}{4L}(1 - \cos 2\alpha) \right] \tag{6.3}$$

where S, α, R and L are slider stroke, crank angle, crank diameter and connecting rod length.

6.3.2 Parameter Change Acquisition

Based on the definition of parameters in the previous section, the acquisition process of parameter change is demonstrated. Through the parameter "ID", parameters can be acquired from the DCMBD models of the original product part and changed product part, and verified to judge whether they are changed. To clarify the process of acquiring the parameter changes, parameter set for a part DCMBD model is presented in Eq. (6.4). Then an algorithm (as shown in Fig. 6.2) is demonstrated to acquire the parameter changes, and the detailed steps are as follows:

$$Set_P(\text{part}\, i) = \{i\_P1, \ldots, i\_Pj, \ldots, i\_Pz\} \quad (6.4)$$

where $Set_P(\text{part}\, i)$ is the name of a parameter set. part *i* is the name of part DCMBD model and *i_Pj* is one of the parameters from part *i*. *z* is the quantity of parameters.

Step1: Initialize *i*=1, *j*=1. // *i, j* are the serial number of //parameters
Step2: Acquire a parameter *k_P i* from the part *k* of original product DCMBD model.
Step3: Acquire a parameter *s_P j* from the part *s* of the changed product DCMBD model.
If *N*(*k_P i*)= *N*(*s_P j*) // *N*(*k_P i*), *N*(*s_P j*) are "ID"s of *k_P i* //and *s_Pj*
Go to ***Step4***
Else
If *j*=*m* // *m* are the quantity of parameters of //part_*s*
Record the "ID" of *k_ Pi* as a deleted parameter in the "LMD" interface
i=*i*+1
Go to ***Step2***
Else
j=*j*+1
Go to ***Step3***
Endif
Endif
Step4: Verification of the *k_P i* and *s_P j*.
If *V*(*k_P i*)≠*V*(*s_P j*) // *V*(*k_P i*), *V*(*s_P j*) are values of *k_P i* //and *s_Pj*
Record the "ID" of *k_ Pi* as a changed parameter in the "LMD" interface
If *i*=*n* // *n* are the quantity of parameters of //part_*k*
Go to ***Step5***
Else
i=*i*+1
Go to ***Step2***
Endif
Endif
Step5: End of program.

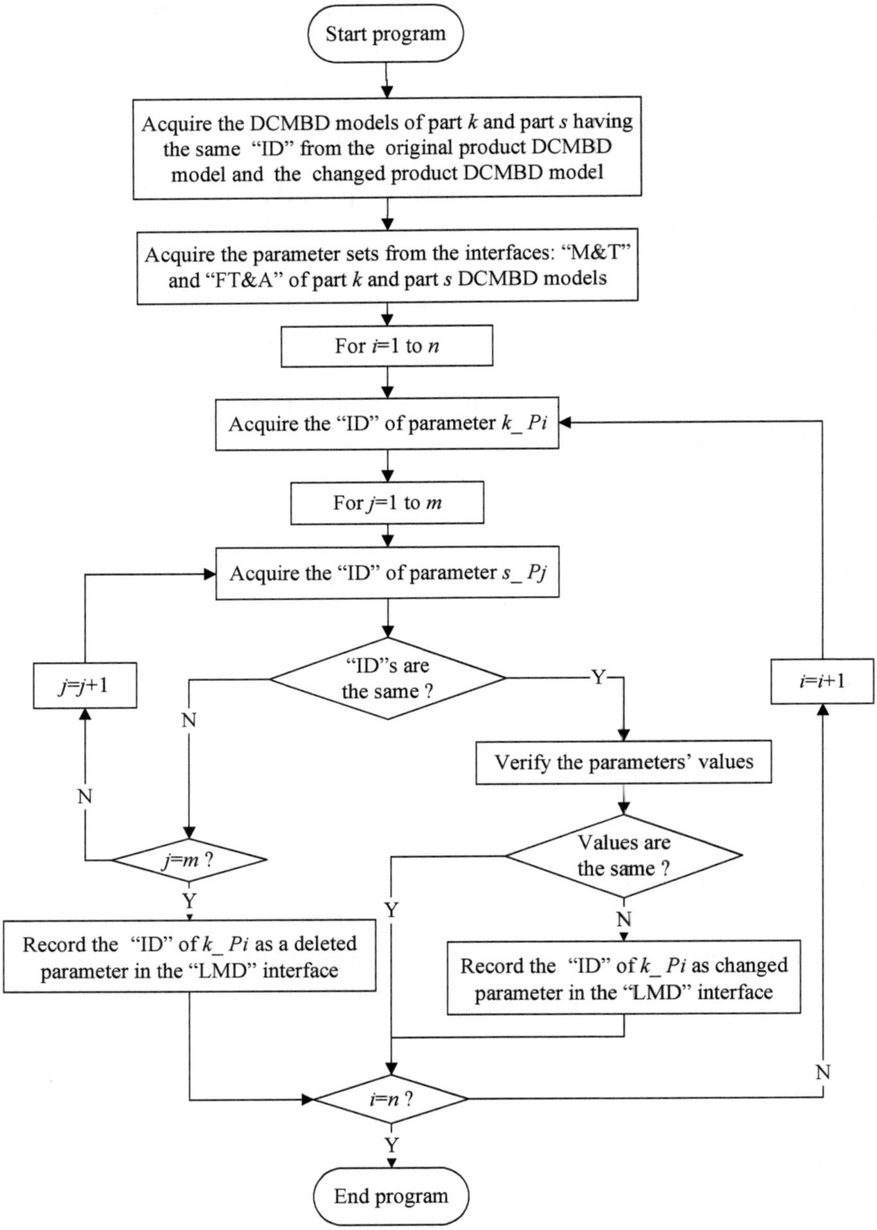

k and *s* are the sequence numbers of part DCMBD models from original product and changed product; *n*,*m* are the quantities of parameters of the two parts respectively.

Fig. 6.2 Flowchart of identifying the parameter change

6.4 Change Acquisition in Terms of Product Assembly

In addition to the parameter change, there is also topology face change in the design change. Parameter change is mainly aimed at addressing design change impact on the design laws and topology face change is mainly aimed at addressing design change impact on the product assembly. In this chapter, the assembly relations between parts are demonstrated through the constraints of topology faces. Thus the changes of topology faces can be applied to describe the impact of design change on product assembly. In this section, first, the concept of part topological decomposition is introduced; second, an algorithm to acquire the topology face change is proposed.

6.4.1 Topological Decomposition of Part

In the product design, semantic feature model can be used for integration of different domains [9]. Meanwhile, topological elements can be used to represent the features of product and describe the geometrical information preciously and completely [10, 11]. In the part modeling, the three-dimensional model is bounded by two-dimensional elements. And the two-dimensional elements are those occupying a certain amount of area. Two-dimensional elements (e.g. faces) are bounded by loops (e.g. circles). Furthermore, a loop is produced by a curve that bends around and crosses itself. In terms of the curve, it is generated by connecting a number of vertices. In this way, a part can be decomposed into five levels of topological elements. For example, a cuboid is shown in Fig. 6.3a, and the corresponding topology structure is shown in Fig. 6.3b, which comprises the decomposition and listing of the topological elements.

The relations of the topological elements in the topology tree can be described in the form of parent–child relations as shown in Table 6.1. For example, the

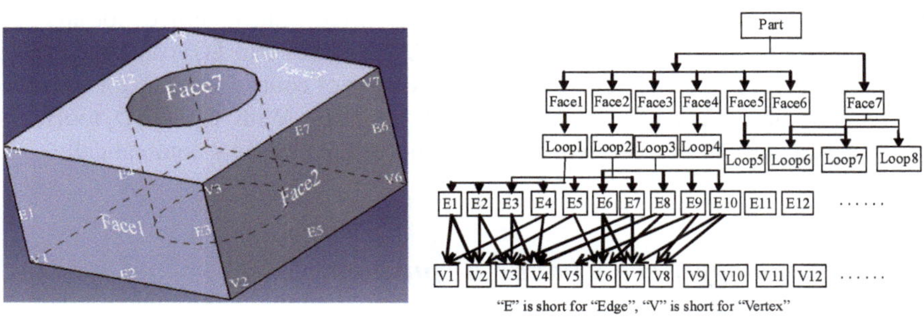

(a) Cuboid with a hole in it (b) Topology tree

Fig. 6.3 Topology structure for a part

Table 6.1 Topological relation between topological elements

Topological relation	Mathematical expression	Example
Vertex-edge	$E(i) = \{E(i) \mid V(j) \in E(i), i \in [1, n], j \in [1, m]\}$ n: quantity of edges of a loop m: quantity of vertices belonging to the edge $E(i)$	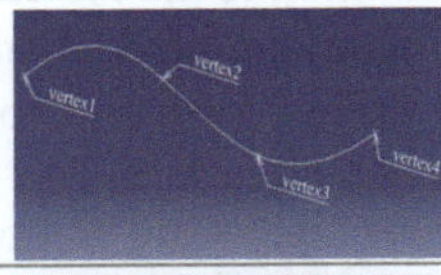
Edge-loop	$L(i) = \{L(i) \mid E(j) \in L(i), i \in [1, n], j \in [1, m]\}$ n: quantity of loops of a face m: quantity of edges belonging to the loop $L(i)$	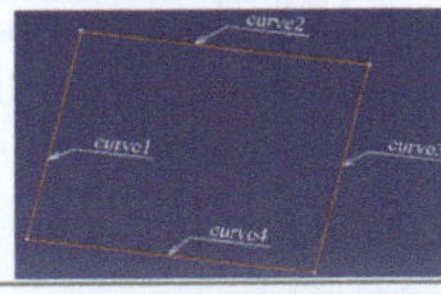
Loop-face	$F(i) = \{F(i) \mid L(j) \in F(i), i \in [1, n], j \in [1, m]\}$ n: quantity of faces of a part m: quantity of loops belonging to the face $F(i)$	
Face-part	$P(i) = \{P(i) \mid F(j) \in P(i), i \in [1, n], j \in [1, m]\}$ n: quantity of parts of a product m: quantity of faces belonging to the part $P(i)$	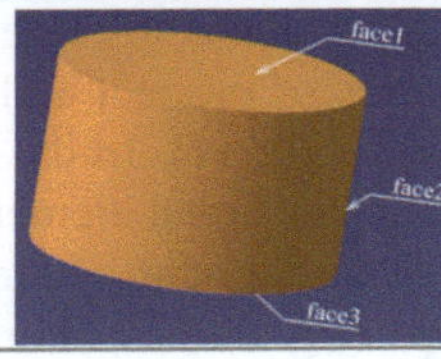

V vertex, *E* edge, *L* loop, *F* face, *P* part

vertex-edge relation indicates that the $E(i)$ is generated through connecting the vertices $V(j)$ ($j = 1, 2, \ldots, m$), which implies that the vertices (named as control vertices) can determine the edge solely. The edge-loop relation indicates that the loop $L(i)$ is generated by connecting the edges $E(j)$ ($j = 1, 2, \ldots, m$). The loop-face and face-part relations can be analyzed similarly.

In addition, the topological elements can be identified through control vertices as shown in Table 6.2. Control vertices are defined as the points acquired from the topological elements in a certain ratio. In terms of the control vertices for an edge, the expression $\{V(1), \ldots, V(i), \ldots, V(n) \rightarrow E(j)\}$ means that $E(j)$ can be identified through the $V(1), \ldots, V(i), \ldots, V(n)$. The expression of control vertices for the other two topological elements can be analyzed similarly. With the CAD tools, the control vertices can be acquired from the topological elements automatically.

6.4.2 Topology Face Change Acquisition

To provide the rationales of topology face change (i.e. the description of topological element changes), changed part model is constructed by modifying the original part

Table 6.2 Control vertices for the topological elements

Type of CV	Mathematical expression	Example
CV for edge	$\{V(1),\ldots,V(i),\ldots,V(n) \rightarrow E(j)\}$ $i \in [1,n], j \in [1,m]$ n: quantity of vertices m: quantity of edges of a loop	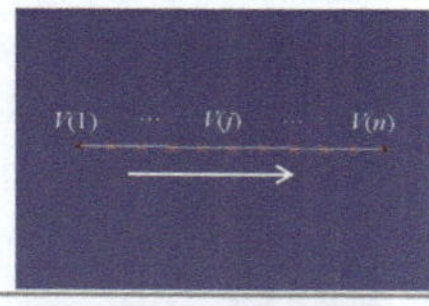
CV for loop	$\{V(1),\ldots,V(i),\ldots,V(n) \rightarrow L(j)\}$ $i \in [1,n], j \in [1,m]$ n: quantity of vertices m: quantity of loops of a face	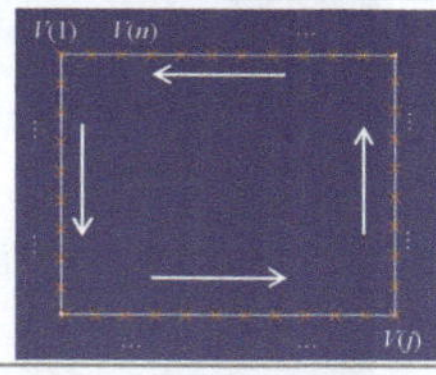
CV for face	$\{V(1),\ldots,V(i),\ldots,V(n) \rightarrow F(j)\}$ $i \in [1,n], j \in [1,m]$ n: quantity of vertices m: quantity of faces of a part	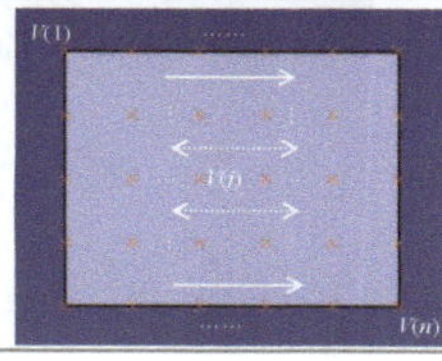

CV control vertices, *V* vertex, *E* edge, *L* loop, *F* face

model in the same CAD environment. Four change types of topology face are concluded: SC (shape change), OC (orientation change), LC (location change), DC (dimension change), and examples are given in Fig. 6.4 to illustrate the change types. To clarify the approach to identify the change type, six properties are listed in Table 6.3. "√" represents that the corresponding property in the column is used to judge the change type in the row.

An algorithm is shown in Fig. 6.5 and detailed as follows:

Step1: Initialize i=1, j=1. // *i, j* are the serial number of topology //faces

Step2: Establish the relations between the part *k* and part *s* having the same "ID"

From the "CG" interface of the part DCMBD model, the topology faces are acquired from the original product part and the changed product part, and they are classified in a one-to-one relation. The process of establishing the relations is demonstrated as follows:

In Fig. 6.6, it is shown that any two topology faces are adjacent to each other when they are connected by common edges. In the schematic diagram, two topological layers are detailed: topology face layer and edge layer. *F*1 is regarded as the basic face and basic face is defined as the first face generated on the first sketch plane in the part DCMBD modeling. And the basic faces from the original part and

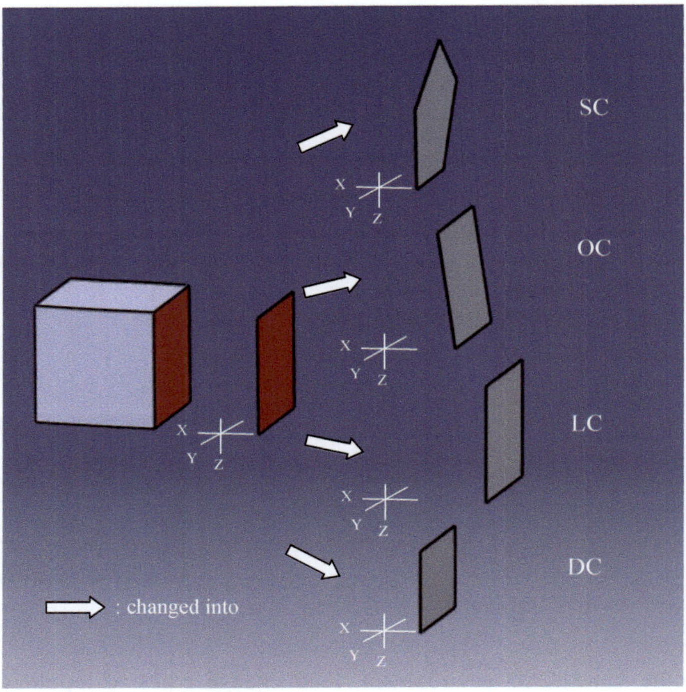

Fig. 6.4 Examples of the change types of topology face

Table 6.3 Properties for the judgment of topology face change type

Change type	Property					
	ID	LQ	EQ	EA	FO	CV
Shape change	√	√	√	√		
Orientation change	√	√	√	√	√	
Location change	√	√	√	√		√
Dimension change	√	√	√	√	√	√

ID identity data, *LQ* loop's quantity, *EQ* edge's quantity, *EA* angle between two edges, *FO* face's orientation, *CV* control vertices

corresponding changed part are established on the same sketch plane. Thus they are linked at the beginning of the part DCMBD modeling. After that, the relations between topology faces from original product part and the changed product part can be acquired through the spatial relations of adjacent topology faces. For example, adjacent relations between an original part faces and adjacent relations between a changed part faces are shown in Fig. 6.7. The basic faces (top faces) of the two parts are both in green in the three-dimensional models and their edges are linked,

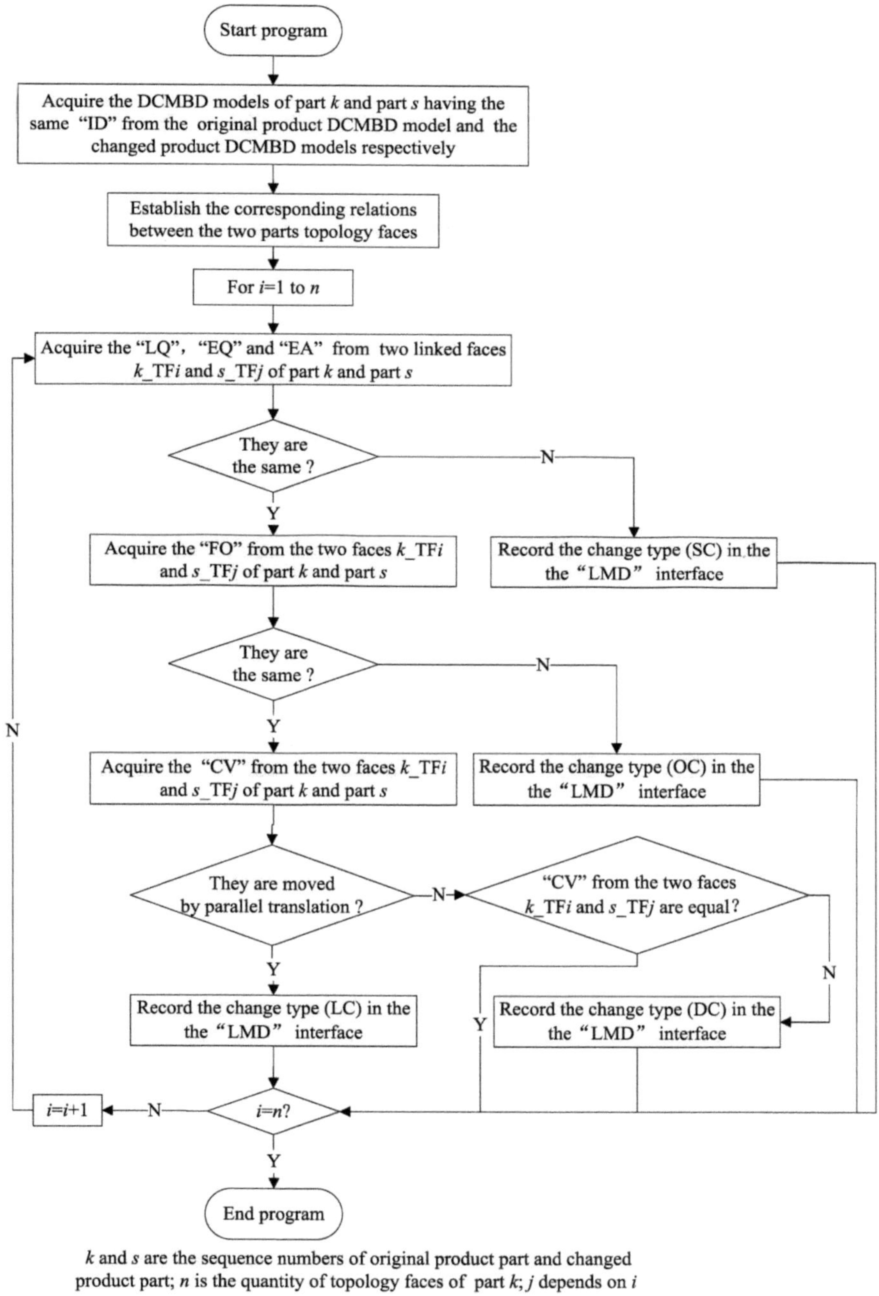

Fig. 6.5 Flowchart of identifying the topology face change

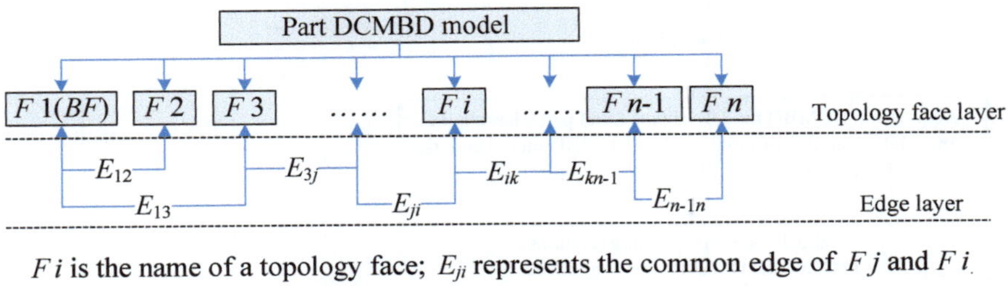

Fig. 6.6 Graph of adjacent faces of a part (a schematic diagram)

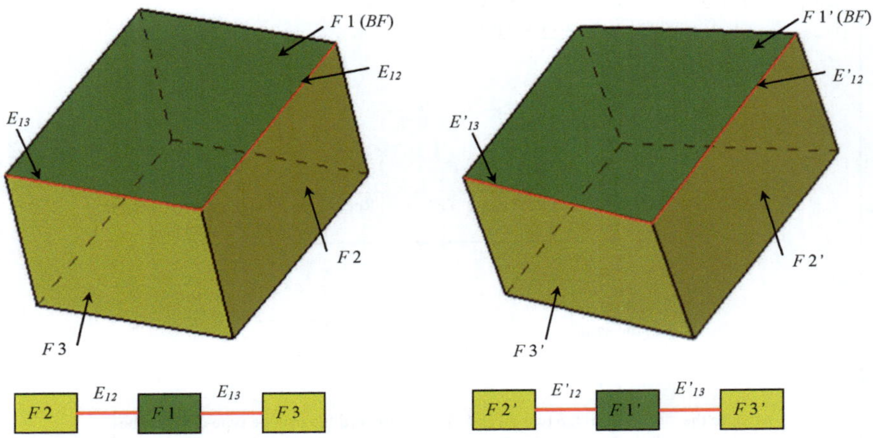

$F1(BF) \cap F2 = E_{12}, F1(BF) \cap F3 = E_{13}$

(a) Topology face adjacent relations for original cuboid

$F'1(BF) \cap F'2 = E'_{12}, F'1(BF) \cap F'3 = E'_{13}$

(b) Topology face adjacent relations for changed cuboid

Fig. 6.7 Example of topology face relations between original and changed parts

respectively, e.g. $E_{12} \leftrightarrow E'_{12}, E_{13} \leftrightarrow E'_{13}$ ("$\leftrightarrow$" represents "is linked to"). Since basic faces and the edges from the original part and changed part are linked, other topology faces can be linked, respectively. That is, $F2$ is linked to $F2'$ and $F3$ is linked to $F3'$, etc.

Step3: Acquire a topology face *k_TFi* from the part *k* of original product DCMBD model.

Step4: Acquire a topology face *s_TFj* linked to *k_TFi* from the part *s* of changed product DCMBD model.

Step5: Verify the "LQ", "EQ" and "EA" of *k_TFi* and *s_TFj*

If *LQ*(*k_TFi*) =*LQ*(*s_TFj*)& *EQ* (*k_TFi*) =*EQ* (*s_TFj*) & *EA* (*E*(*a*), *E*(*b*)) =*EA* (*E*(*a'*), *E*(*b'*)) // *LQ* (*X*), *EQ* (*X*), *EA* (*Y*,*Z*) are quantity of loop, edge, and edge's angle

Go to ***Step6***

Else

Record the change type of *k_ TFi* as "SC" in the "LMD" interface and Go to ***Step8***

Endif

Step6: Verify the "FO" of *k_TFi* and *s_TFj*

If *FO*(*k_TFi*) =*FO*(*s_TFj*) // *FO*(*X*) is *face's* direction //vector

Go to ***Step7***

Else

Record the change type of *k_ TFi* as "OC" in the "LMD" interface and Go to ***Step8***

Endif

Step7: Verify the "CV" of *k_TFi* and *s_TFj*

If *C*(*V*(*p*))—*C* (*V*(*p'*)) =*C*(*V*(*q*))—*C* (*V*(*q'*))≠0 //*C*(*X*) is vertex's //coordinate value

Record the change type of *k_ TFi* as "LC" in the "LMD" interface and Go to ***Step8***

Else

If *C*(*V*(*p*)) —*C* (*V*(*p'*)) =0

Go to ***Step8***

Else

Record the change type of *k_ TFi* as "DC" in the "LMD" interface and Go to ***Step8***

Endif

Endif

Step8: Judge whether the process is ended.

If *i*=n // *n* is the quantity of topology faces //in part *k*

Go to ***Step9***

Else

i=*i*+1

Go to ***Step3***

Step9: End of program. Based on the proposed algorithm, the initial changes of topology faces can be acquired, and stored in the part DCMBD model and database automatically. Then initial changes of topology faces can be propagated with the product relation models.

6.5 Relation Model for MBD-Based Change Propagation Analysis

6.5.1 Parameter Relation

Product design can be regarded as a process of determining the parameters and constructing the relations between them according to the design laws. The modular diagram of parameter relation acquisition is shown in Fig. 6.8 (outputs of each stage are shown in boxes with dashed arrows).

The parameter relation is recorded in the "R" interface during the product DCMBD modeling. A data interpreter deals with the information of the product DCMBD models in the form of a hierarchical frame as follows:

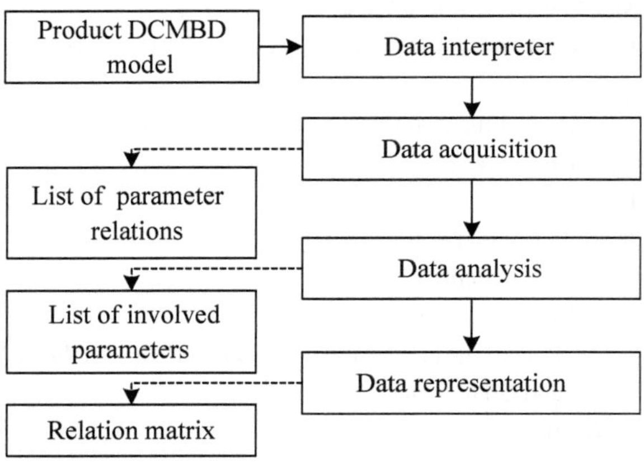

Fig. 6.8 Parameter relation acquisition

product = Product_name
composed_of = Part 1, *Part* 2, ...
 *part = Part*_name (*Part* 1)
 part_of = Product_name
 annotated_by =Part 1*_parameter* 1, *Part* 1*_parameter* 2, ...
  parameter = Part 1*_parameter* 1
  value = value_of_ Part 1*_parameter* 1
  linked_with = Part i_parameter j, ...
  parameter =Part 1*_parameter* 2
  value = value_of_ Part 1*_parameter* 2
  linked_with = Part l_parameter m, ...
  ... ...

The *Product_name* is used to identify a product, and the *composed_of* lists the parts that constitute the product. *Part_name* is the name of the parent part and *annotated_by* lists the corresponding parameters' names. *value* determines the parameter's values. Besides, the *linked_with* records the related parameters' "IDs". For example, the *Product_name* can be "press machine". *composed_of* lists its parts: slider, rod, crank, frame. *Part_name* can be "connecting rod". *annotated_by* lists the parameters of the connecting rod: "length", "diameter", "material", etc. The *value* can be the "10 cm" for the "length". The *linked_with* lists the related parameters of the "length", e.g. "diameter" of the crank, "stroke" of the slider, etc.

The data acquisition module parses the frame to acquire the parameter relations and stores them in the database. The output of the data acquisition module is a list of relations between parameters. The general form of parameter relations is as follow: *Relation [exogenous/constrained parameter (parameter_name), constrained/exogenous parameter (parameter_name), assignment_flag]*.

parameter_name is applied to retrieve the corresponding parameter. The value of *assignment_flag* records the symbol of the parameter relation. For example, the relation between "length" of rod and "stroke" of the slider can be described as *Relation [2_L, 1_S, 2_R1]*, where *2_L* is the "ID" of "length" of rod; *1_S* is the "ID" of "stroke" of slider; *2_R1* represents the symbol of this parameter relation.

The data analysis module parses the hierarchical frame to search for the parameters and retrieves corresponding parent part. The general form of the data analysis module's output is as follow:

[part (part_name), exogenous parameter (parameter_name), exogenous parameter (parameter_name),..., constrained parameter (parameter_name), constrained parameter (parameter_name),...]

The data representation module arranges the above data in the form of PRM (parameter relation matrix). In the PRM, a mark in a cell indicates that there are one or more relations between the initiating part in the row and the affected part in the column. An example is given in Fig. 6.9. "#" means that the part from the row can affect the part from the column. And each relation is associated with an information

Affected parts

Initiating parts

	Part 1	*Part* 2	……	*Part m*-1	*Part m*
Part 1	#		……	#	
Part 2	#		……	#	#
⋮	⋮	⋮	⋮	⋮	⋮
Part m-1		#	……	#	
Part m	#	#	……		#

Fig. 6.9 Parameter relation between parts DCMBD models

Affected parts

Initiating parts

	slider	rod	crank	frame
slider		#	#	#
rod	#	#	#	
crank	#	#	#	#
frame	#		#	

ID	Content	Description
2_R1	*1_S* & *2_L*	The stroke of slider can affect the length of the rod.

1_S: stroke of slider; *2_L*: length of rod

Fig. 6.10 Example of parameter relations for a press machine

field that details the parameters of initiating and affected parts. For the example of press machine, the parameter matrix can be demonstrated in Fig. 6.10.

6.5.2 Topology Face Relation

Product is considered to be assembled through mutual constraints between parts topology faces. Besides, adjacent faces within a part are connected through common edges. Therefore, the topology face relation is divided into topology face relation between parts and topology face relation within a part.

6.5.2.1 Topology Face Relation Between Parts

The modular diagram of acquiring the topology face relation between parts is shown in Fig. 6.11 (outputs of each stage are shown in boxes with dashed arrows).

The topology face relations are stored in the "GC" interface of the product DCMBD models during the product modeling process. A data interpreter parses

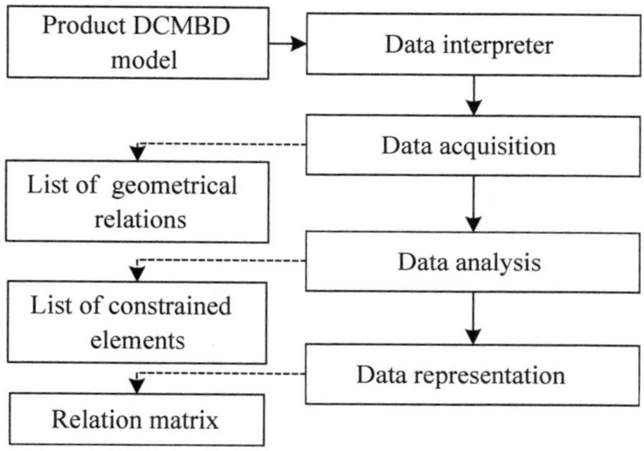

Fig. 6.11 Acquisition of topology face relation between parts

and stores the information of the product DCMBD models in the form of a hierarchical frame as follows:

product = *Product_name*
composed_of = *Part* 1, *Part* 2, ...
  part = *Part_name* (*Part* 1)
  part_of = *Product_name*
  bounded_by = *Part* 1*_face* 1, *Part* 1*_face* 2, ...
    face = *Part* 1*_face* 1
    control_vertices = *Part* 1*_face* 1*_control_vertices*
    constrained_with = *Part i_face j*
    face = *Part* 1*_face* 2
    control_vertices = *Part* 1*_face* 2*_control_vertices*
    constrained_with = *Part t_face m*
    ……

The *Product_name* is used to identify a product, and the *composed_of* lists the parts that constitute the product. *Part_name* is the name of the parent part and *bounded_by* lists the topology faces bounding the part. The *constrained_with* lists the faces of other parts that are related to the faces of the parent part. Besides, the *control_vertices* is used to identify the topology face. For example, the *Product_name* can be "press machine". *composed_of* can list the parts constituting the press machine. *Part_name* can be "connecting rod". The *bounded_by* lists the topology faces bounding the connecting rod in terms of the control vertices. *constrained_with* lists the topology faces of other part constrained with those of connecting rod.

The data acquisition module parses the frame of product DCMBD models to acquire the topology face relations. During this stage, the topology face relations from the "GC" interface along with faces' control vertices are acquired. The general form of topology face relation is as follow:

Constraint [face (control vertices), face (control vertices), assignment_flag]

The *control vertices* are used to retrieve the constrained faces. The value of *assignment_flag* records the symbol of the topology face relation. For example, constraint between the inner cylindrical surface of rod and the outer cylindrical surface of crank can be described as *Constraint [2_F1, 3_F3, 2_GC1]*, where *2_F1* is the inner cylindrical surface of rod; *3_F3* is the outer cylindrical surface of crank; *2_GC1* represents the symbol of this topology face constraint.

The data analysis module uses the control vertices to output a list of constrained elements. The list of constrained elements includes constrained faces and corresponding parent part. The general form of constrained elements is:

[part (part name), face (control vertices), assignment_flag]

The *control vertices* determine the corresponding face and further identify the corresponding parent part. The value of *assignment_flag* records the symbol of topology face relation, in which the face is involved. Taking the inner cylindrical surface of rod as an example, the form of constrained element can be expressed as *[rod, 2_F1, 2_GC1]*.

The data representation module arranges the constraints between parts in the form of TFRM (topology face relation matrix) as shown in Fig. 6.12. In the TFRM, a mark in a cell indicates that there are one or more constraints between the initiating part in the row and the affected part in the column. Besides, a general form of the *RM (Part l, Part t)* is shown in Eq. (6.5). For example, the relation model of press machine can be demonstrated in Fig. 6.13.

$$RM(Part\,l, Part\,t) = \begin{pmatrix} a_{11} & \cdots & a_{1j} & \cdots & a_{1n} \\ \vdots & & \ddots & & \vdots \\ a_{i1} & \cdots & a_{ij} & \cdots & a_{in} \\ \vdots & & \ddots & & \vdots \\ a_{m1} & \cdots & a_{mj} & \cdots & a_{mn} \end{pmatrix} \tag{6.5}$$

Fig. 6.12 Constraints between parts DCMBD models formatted in RM

Affected parts (columns); Initiating parts (rows)

	Part 1	*Part* 2	……	*Partm-1*	*Partm*
Part 1		#	……		
Part 2	#		……	#	
⋮	⋮	⋮		⋮	⋮
Partm-1		#	……		#
Partm			……	#	

Affected parts (columns) / Initiating parts (rows)

	slider	rod	crank	frame
slider		#		#
rod	#		#	
crank		#		#
frame	#		#	

(a) RM of the press machine

crank (columns) / rod (rows)

	F_1	F_2	F_3	F_4	F_5	F_6
F_1	0	0	1	0	0	0
F_2	0	0	0	0	0	0
F_3	0	0	0	1	0	0
F_4	0	0	0	0	1	0
F_5	0	0	0	0	0	0

(b) Part-part *RM* (rod, crank)

Fig. 6.13 Example of topology face relations for a press machine

where *Part l* and *Part t* are the names of two constrained parts. a_{ij} represents the topology face relation between the *i*th face of *Part l* and the *j*th face of *Part t*. a_{ij} can be valued as 1 or 0. 1 (or 0) means that the two faces from the parts are (or are not) constrained. *m* and *n* are the quantity of faces of *Part l* and *Part t*.

6.5.2.2 Topology Face Relation Within a Part

The acquisition of topology face relation within a part is similar to that of topology face relation between parts. The difference is that the topology face relation within a part is acquired from the "CG" interface of the part DCMBD model. The modular diagram of acquiring the topology face relation within a part is shown in Fig. 6.14 (outputs of each stage are shown in boxes with dashed arrows).

A data interpreter parses the part DCMBD model and stores the information of the part DCMBD model in the form of a hierarchical frame as follows:

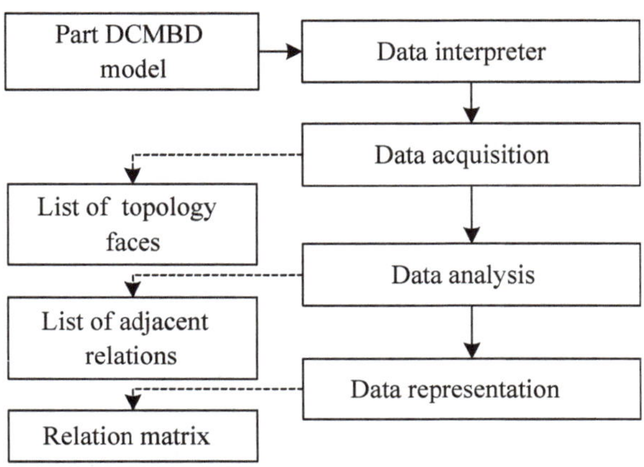

Fig. 6.14 Acquisition of topology face relation within a part DCMBD model

part = *Part_name*
bounded_by = *Part_name_face* 1, *Part_name_face* 2, ...
face = *Part_name_face* 1
control_vertices = *Part_name_face* 1*_control_vertices*
adjacent_to = *Part_name_face i*, *Part_name_face j*, ...
face = *Part_name_face* 2
control_vertices= *Part_name_face* 2*_control_vertices*
adjacent_to = *Part_name_face t*, *Part_name_face m*, ...
… …

Part_name is the name of the parent part and *bounded_by* lists the faces that constitute the part. The *adjacent_to* lists the faces that are adjacent to one specific face from the parent part. Besides, the *control_vertices* identify the topology face. For example, the *Part_name* can be "connecting rod" of a press machine. The attribute *bounded_by* lists the topology faces bounding the connecting rod in terms of the control vertices. *adjacent_to* lists the adjacent relation between topology faces, e.g. *2_F1*, *2_F2* of the rod are adjacent to each other as they have common edges.

During the stage of data acquisition, the adjacent relations between the identified faces are acquired. The output of the data acquisition module is a list of adjacent relations between topology faces as follow:

Adjacency [face (control vertices), face (control vertices), assignment_flag]

The *control vertices* are used to retrieve the corresponding topology face. *assignment_flag* records the symbol of the adjacent relation between the faces. For example, adjacency between the topology face *2_F1* and *2_F2* can be described as *Adjacency [2_F1, 2_F2, 2_AD1]*, where *2_F1*, *2_F2* are two adjacent faces; *2_AD1* represents the symbol of this adjacent relation.

The data analysis module parses the frame of part DCMBD model to acquire all the faces in the part though their control vertices. The output of the data analysis module is a list of faces as follow:

[part (Part_name), face (control vertices), face (control vertices), ...]

The data representation module arranges the above data in the form of part RM as shown in Fig. 6.15. A general form of the part RM *RM* (*Part l*) is shown in Eq. (6.6). Taking connecting rod of press machine as an example, the relation model *RM* (rod) can be demonstrated in Eq. (6.7).

$$RM(Part\,l) = \begin{pmatrix} a_{11} & \cdots & a_{1j} & 1 & a_{1n} \\ \vdots & & \ddots & & \vdots \\ a_{i1} & \cdots & a_{ij} & \cdots & a_{in} \\ \vdots & & \ddots & & \vdots \\ a_{n1} & \cdots & a_{nj} & \cdots & a_{nn} \end{pmatrix} \tag{6.6}$$

where *Part l* is the name of a part DCMBD model. a_{ij} represents the adjacent relation between the *i*th face and the *j*th face of *Part l*. a_{ij} can be valued as 1 or 0.

Fig. 6.15 Adjacent relations within a part DCMBD model formatted in RM

Part *l*

Part *l*	*F* 1	*F* 2	……	*F m-1*	*F m*
F 1	a_{11}	a_{12}	……	$a_{1m\text{-}1}$	a_{1m}
F 2	a_{21}	a_{22}	……	$a_{2m\text{-}1}$	a_{2m}
⋮	⋮	⋮		⋮	⋮
F m-1	$a_{m\text{-}11}$	$a_{m\text{-}12}$	……	$a_{m\text{-}1m\text{-}1}$	$a_{m\text{-}1m}$
F m	a_{m1}	a_{m2}	……	$a_{mm\text{-}1}$	a_{mm}

1 (or 0) means that the two faces have (or have no) common edges with each other. n is the quantity of topology faces of *Part l.*

$$RM(rod) = \begin{pmatrix} 0 & 1 & 0 & 1 & 0 \\ 1 & 0 & 1 & 0 & 1 \\ 0 & 1 & 0 & 1 & 0 \\ 1 & 0 & 1 & 0 & 1 \\ 0 & 1 & 0 & 1 & 0 \end{pmatrix} \tag{6.7}$$

6.6 Change Propagation Analysis

Change propagation can result in further changes to the product design, which may cause severe profit losses for companies. Thus, design change solutions are proposed not only to meet the initial changed requirement of performance index, but also to overcome the impacts of change propagation. Two types of change propagation: parameter propagation and topology face propagation are introduced to realize these targets, which are aimed at addressing design change impact on the design laws and product assembly, respectively. Besides, they correspond to different phases of product design change. That is, in the first phase, new requirement of product performance index is raised. For example, long slider stroke of the press machine (discussed in Sect. 4.1) is required. Then parameters of other parts in the Eq. (6.3) may be affected and changed to meet the corresponding design laws. In the next phase, topology face propagation of affected parts is implemented to meet assembly requirements of the affected parts.

6.6.1 *Type of Change Propagation*

Two types of change propagation are as follows:

1. Parameter propagation: when parameter changes are proposed to meet the changed requirements of product performance index, other related parameters can be affected through the parameter relations. This process is called parameter propagation.
2. Topology face propagation: after parameters of parts are changed, the geometrical structure of changed parts may also be changed, which can influence the assembly of the changed product. That is, the changed parts can propagate topology face changes to other parts through the topology face relations. This process is called topology face propagation.

6.6.2 *Change Propagation Pattern*

Two change propagation patterns are proposed with respect to the two phases of change propagation.

6.6.2.1 Parameter Propagation Pattern

In the product design change, first, constrained parameter of product is changed to satisfy the changed requirement of product performance index. Then, the parameter change will be propagated to exogenous parameters due to the parameter relations. After that, changes of exogenous parameters may be propagated to other constrained parameters. The process of parameter change propagation will end when the last affected exogenous parameter cannot propagate changes to other constrained parameters. During this process, an index is introduced to describe the impact of parameter propagation: parameter change influence (*PCI*).

PCI is defined as the quantity of affected parameters with respect to each change propagation path. The affected parameter quantity of the *i*th change propagation path is named as *PCI(i)*. In the design change, a recommendation is that smaller *PCI(i)* should be given priority as it is beneficial for the reduction of influence scope.

6.6.2.2 Topology Face Propagation Pattern

When changed parameters satisfy the changed requirement, the geometrical structure of the changed parts may be changed, which will propagate changes to other constrained parts due to topology face relations. Change vector (CV) is proposed to describe the changes of a part's topology faces as shown in Eq. (6.8). Besides, a formula is applied to describe the topology face change propagation in Eq. (6.9). The change propagation process will not end until the change vector of the last affected part is empty.

$$CV(Part\ l) = \left(\begin{pmatrix} a_1 \\ \chi \end{pmatrix}, \ldots, \begin{pmatrix} a_i \\ \chi \end{pmatrix}, \ldots \begin{pmatrix} a_n \\ \chi \end{pmatrix}\right) \tag{6.8}$$

where *Part l* is the name of a changed part. $\begin{pmatrix} a_i \\ \chi \end{pmatrix}$ is the change state of the *i*th topology face. a_i can be valued as 1 (or 0), which means the corresponding topology face is changed (or not changed). χ is change type of corresponding topology face and can be "SC", "OC", "LC" or "DC". n is the quantity of topology faces of *Part l*.

$$CV(Part\,k) = \prod_{i=j}^{k} CV(Part\,i) * RM(Part\,i, Part\,i+1) * RM(Part\,i+1), \quad k > j \tag{6.9}$$

where i is the serial number of influenced part along a change propagation path. j is the serial number of an upstream changed part. k is the serial number of the affected part along the change propagation path.

During the topology face propagation, topology face change influence (*TFCI*) index is introduced to describe the impact of topology face propagation. Quantity of affected topology faces in the *i*th change propagation path is named as *TFCI (i)*. *TFCI* indicates the impact degree of change propagation and smaller *TFCI* is favorable.

6.6.3 Design Change Process

As described previously, product design change process is assumed to be composed of two repeated phases: design change of parameters and design change of topology faces. A flowchart is shown in Fig. 6.16 to demonstrate the design change process and details are as follows:

Step1: Changed requirements of product performance index are analyzed by the designers to modify the corresponding constrained parameters, e.g. improving the output power of a transmission platform.

Step2: The change of constrained (affected) parameter is propagated through checking the parameter relation matrix with the pattern described in Section 6.6.2.1. In each propagation step, implement the following judgment:

If no parameters are affected throughout the whole propagation process
    Go to *Step11*
Else
    If no parameters are affected in this single step
        Go to *Step3*
    Else
        Go to *Step2*
    Endif
Endif

Step3: The index *PCI* for each of the change propagation paths (as candidates of design change solution) is calculated.

Step4: Based on the possible propagation paths and the calculated *PCI*, discussion can be made among related stakeholders (customers, financial executives, designers, etc). Then the optimal solution will be selected (reselected) and validated. After that, the affected parts DCMBD models are changed by the designers.

Step5: Verify the changed part DCMBD models with the corresponding original parts DCMBD models to acquire the changes of topology faces with the method described in Section 6.4.2.

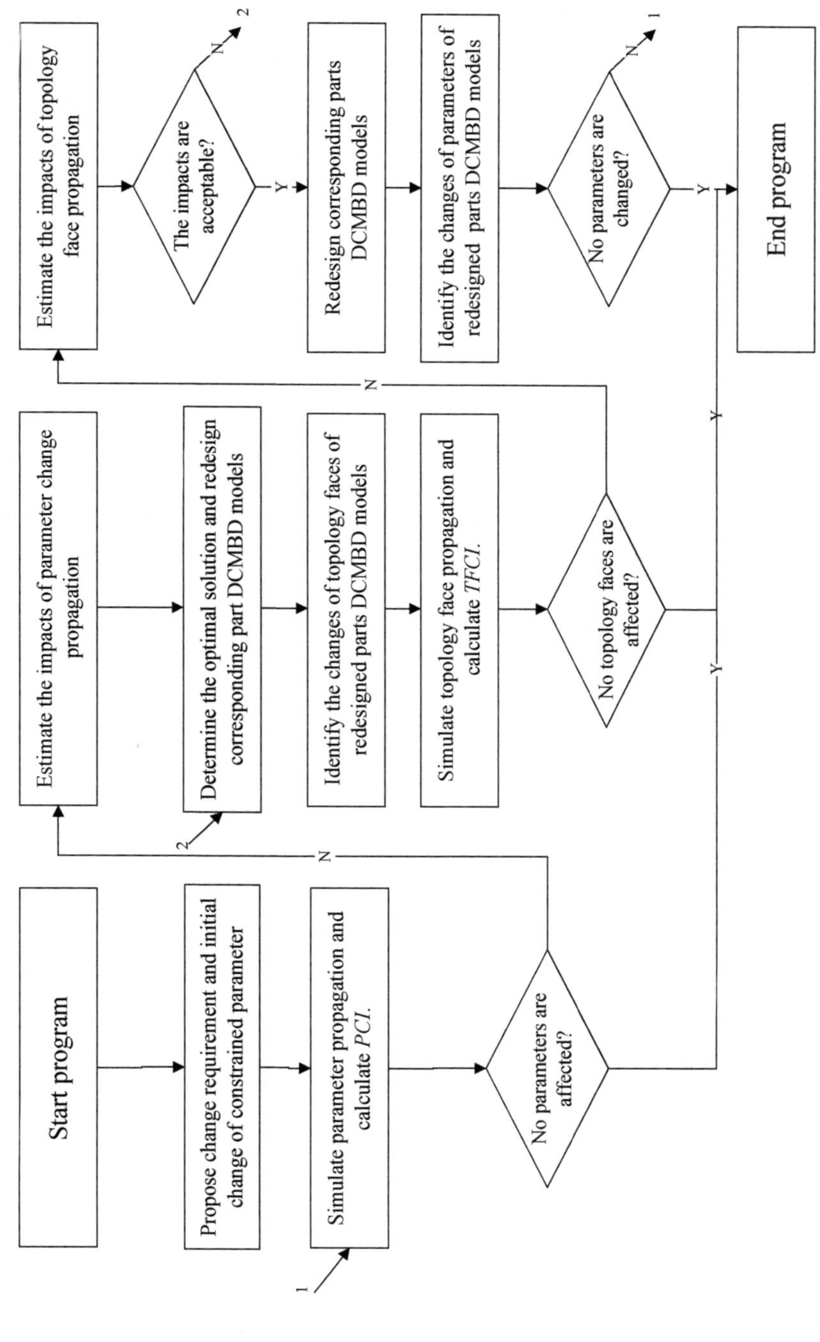

Fig. 6.16 Flowchart of design change process

Step6: The initiating (affected) changes of topology faces are propagated to other parts with the pattern described in Section 6.6.2.2. In each propagation step, implement the following judgment:

If no topology faces are affected throughout the whole propagation process
  Go to *Step11*
Else
  If no topology faces are affected in this single step
    Go to *Step7*
  Else
    Go to *Step6*
  Endif
Endif

Step7: The index *TFCI* for each of possible propagation paths (as candidates of design change solution) is calculated.

Step8: Based on the possible propagation paths and the calculated *TFCI*, discussion can be made among related stakeholders. Then implement the following judgment:

If impacts of the affected parts DCMBD models are acceptable
  Go to *Step9*
Else
  Go to *Step4*
Endif

Step9: The affected parts DCMBD models are redesigned by designers to construct changed parts DCMBD models.

Step10: Verify the changed parts DCMBD models with the corresponding original parts DCMBD models to identify the parameter changes with the method described in Section 6.3.2. Then implement the following judgment:

If no parameters are changed
  Go to *Step11*
Else
  Go to *Step2*
Endif

Step11: End of program.

6.7 Application of MBD-Based Change Management

Based on the developed method and algorithms, a computer-aided design change system is developed with CATIA V5R18 and Microsoft Office Access 2007, including the module of acquiring changes, module of acquiring relation matrices, module of implementing change propagation and module of outputting the change propagation results. CATIA V5R18 is used to construct the product DCMBD models and Microsoft Office Access 2007 is applied to record the data of the product DCMBD models. Using the Visual C^{++} and CAA (component application architecture) technology, an interface is developed to interpret the product DCMBD models and store the interpreted data in the created datasheets. Material-transfer platform can be regarded as a mature industrial product, and its design process is quite complete and familiar to engineers. In this chapter, case study of a material-transfer platform was carried out to show how the developed method can help the designers in the design change process. The product was simplified and the main parts (modules) were constructed to demonstrate the initial evaluation of the method. The main parts of the product are listed in Table 6.4 and its DCMBD models are shown in Fig. 6.17.

The transmission speed of the existing device was slow and cannot meet the new application requirement. It was necessary to modify the material-transfer platform to increase the transmission speed. Since the transmission mechanism was the critical unit of the material-transfer platform, it was required that the transmission components of the existing device be used as far as possible.

To achieve the higher transmission speed, the originating constrained parameter 14_V (speed of the transmission belt) was selected to be changed. Then its change was propagated to other parts with the parameter change propagation pattern described in Sect. 6.6.2.1. With the help of prototype system, the change

Table 6.4 Main parts of material-transfer platform

No.	Part name
1	Electromotor
2	Coupling1
3	Bearing1
4	Bearing2
5	Shaft1
6	Gear1
7	Bearing3
8	Bearing4
9	Shaft2
10	Gear2
11	Chassis
12	Coupling2
13	Roller
14	Transmission belt

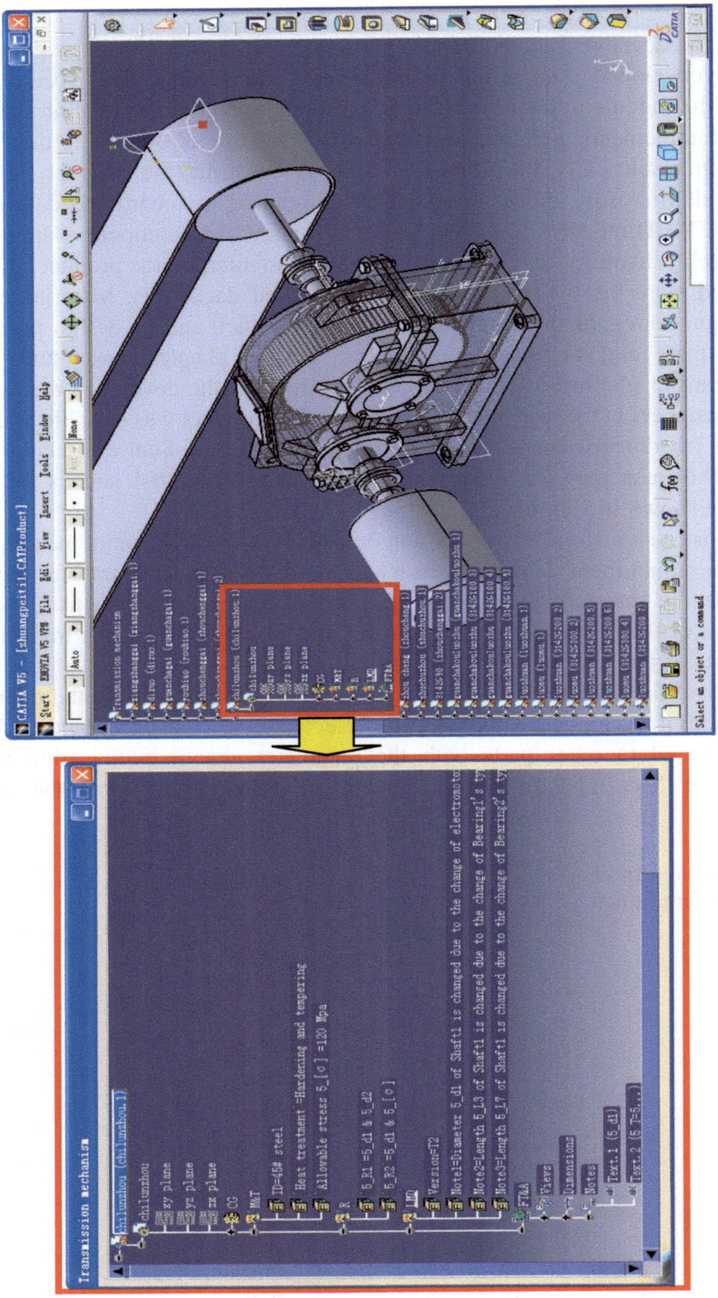

Fig. 6.17 DCMBD models of material-transfer platform

propagation paths and the details of each change propagation path can be acquired. In Fig. 6.18, possible change propagation paths are acquired and listed. The affected parameters of any selected propagation path can be listed and the annotations on the CAD models are marked with red color by the prototype system. For example, the affected parameters of the second change propagation path are listed and marked with red color as shown in Fig. 6.19. 14_V, 1_Pd, 1_Nd, 5_T, 5_d1, 3_Type, 4_Type represent parameters: speed of transmission belt, electromotor power, electromotor speed, torque of shaft1, diameter of shaft1, type of bearing1 and type of bearing2. *PCI* index of this change propagation path is 7. Based on the above analysis, the change impacts were assessed by designers and meetings were held among the stakeholders to select an optimal design change solution. Moreover, comments can be continuously fed back to modify the design change solution if the impacts of design change are unacceptable by the stakeholders. In this case, considering the cost and time of the design change, the second propagation path (14–1–5–3, 4, i.e. transmission belt → electromotor → shaft1 → bearing1, bearing2) was chosen as the design change solution and the related parameters were modified by the designers.

After that, the redesigned parts DCMBD models were verified with the corresponding original parts DCMBD models to acquire the topology face changes. Then the topology face changes were propagated to other parts with the topology face change propagation pattern described in Sect. 6.6.2.2. The change propagation

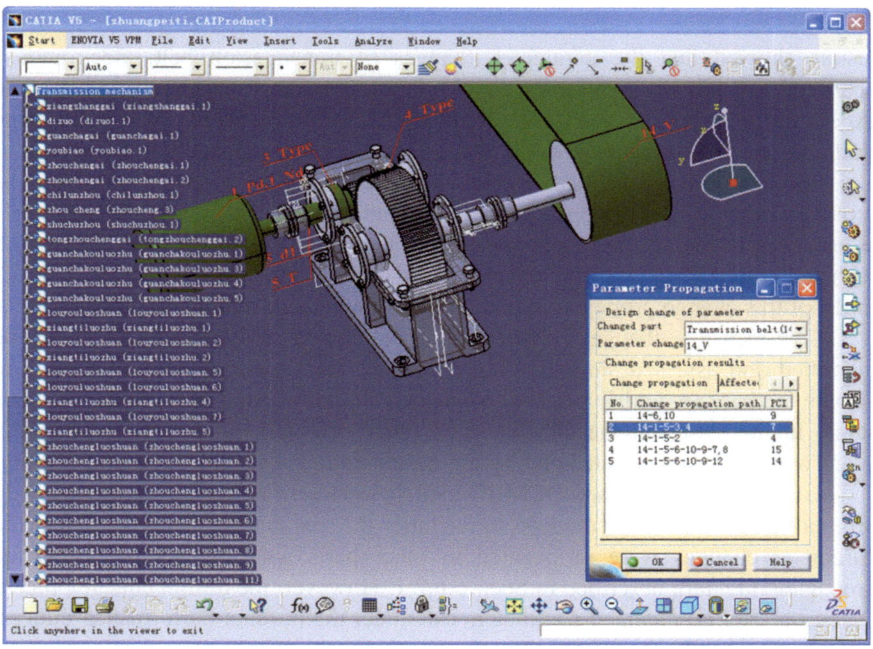

Fig. 6.18 Parameter change propagation

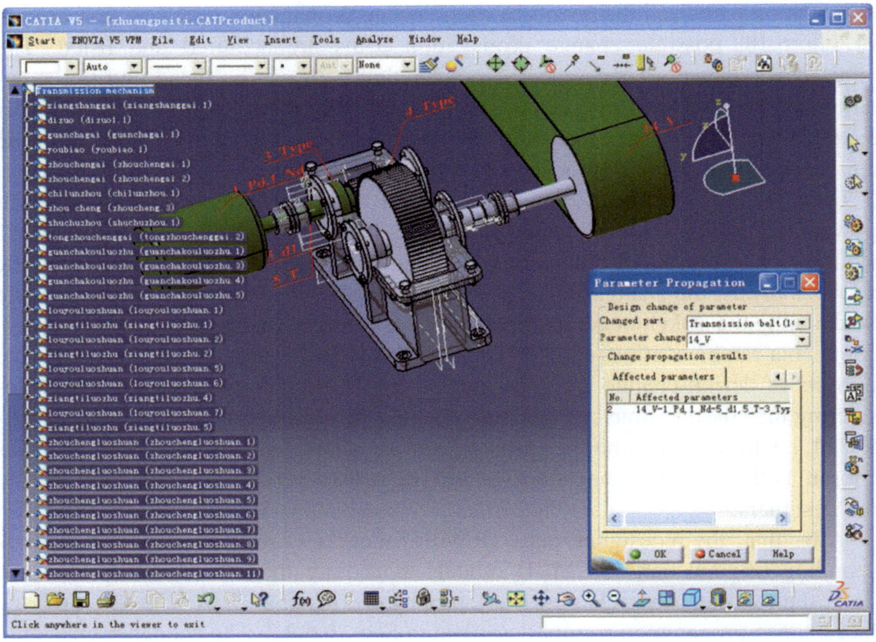

Fig. 6.19 Analysis of parameter change propagation

results showed that the parameter changes of transmission belt, electromotor, bearring1 and bearing2 caused no topology face change propagation to other parts, except the changes of shaft1. The topology face change propagation paths from shaft1 are acquired and listed as shown in Fig. 6.20. The affected topology faces of any selected propagation path can be listed and marked with red color by the prototype system. Taking one change propagation path (5–3, i.e. shaft1-bearing1) as an example, the affected topology faces of the parts DCMBD models are marked with red color to highlight the changes for designers as shown in Fig. 6.21. *TFCI* index of this change propagation path is 4 and the change type of the affected faces is DC (dimension change). Impacts of other change propagation paths can be identified and assessed similarly. If the impacts are unacceptable, comments can also be fed back to modify the design change solution of parameters in the previous stage. In this case, all the affected parts (shaft1, coupling1, bearing1, bearing2, gear1 and chassis) were changed by the designers to satisfy the product assembly.

Next, the redesigned parts DCMBD models were verified with the corresponding original parts DCMBD models to acquire the parameter changes. Then the parameter changes were propagated to other parts with the parameter change propagation pattern described in Sect. 6.6.2.1. However, the design change process ended because no changed parameters can affect the product performance indices.

It is known that the traditional engineering change management mainly relies on experienced designers for the product change-related data acquisition and change

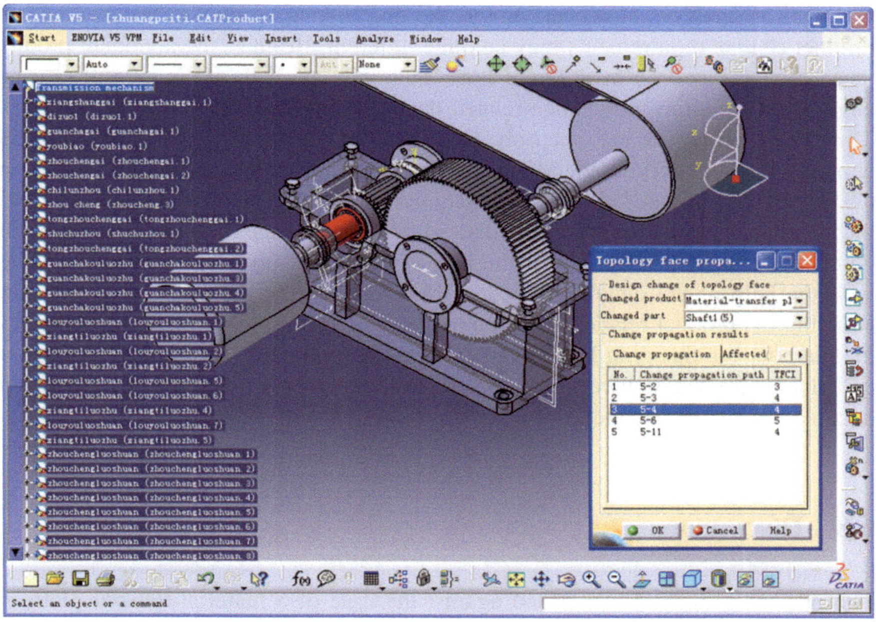

Fig. 6.20 Topology face change propagation

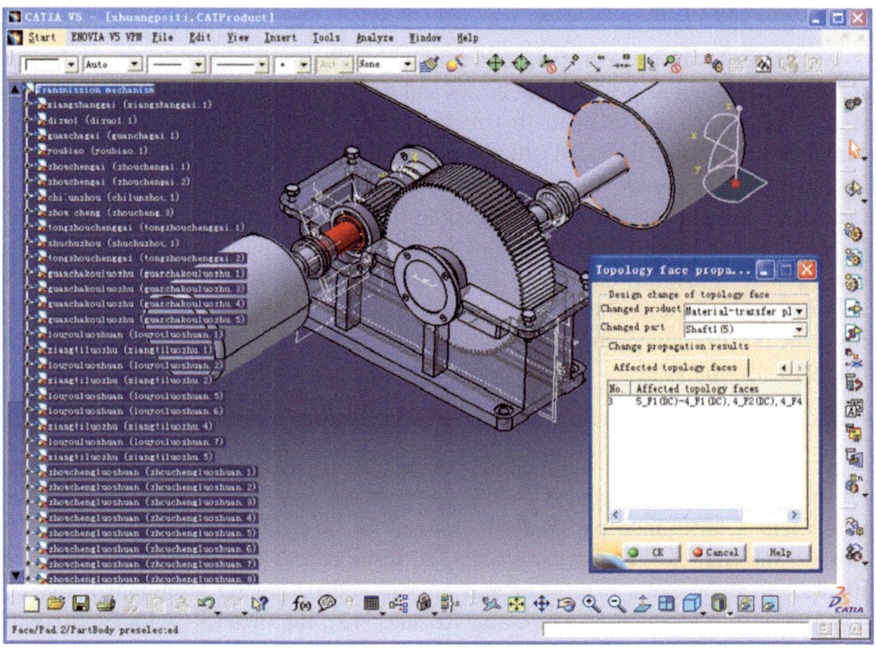

Fig. 6.21 Analysis of topology face change propagation

propagation analysis. However, due to the diversity in product customization, the product change requirements may be quite different. Besides, the abundant product data is usually scatted in different files. This will result in much pressure on the engineers to acquire the change-related data and analyze the change propagation. Thus these processes are not only time-consuming but also error-prone. In the above case study, without the MBD-based method and developed prototype system, designers spent about half a day to acquire and propagate changes based on the technical files, and then store the data of product changes into the database. By comparison, it took about 2 h if the above mentioned processes were carried out with the help of the developed system (1 h for defining the parameters and constructing the product relations, half an hour for propagating and analyzing changes, half an hour for changing DCMBD models). This is because the change acquisition and change propagation can be automatically implemented, and the change-related data is acquired automatically from the product DCMBD models.

6.8 Discussions

The presented method was tested against the case of a material-transfer platform and evaluated initially. In this section, further discussions on the developed method are given.

- Complexity of the product

As discussed in the Sect. 6.3, modern CAD software is able to manage the data of DCMBD model as geometrical entity and the DCMBD model is the prerequisite of the change management in this method. It is an effort-consuming process to establish the product DCMBD model when the complexity of product increases. One approach to tackle this issue is to develop a definition module to aid the designers to record the product/part data in the DCMBD model automatically, and the designers can update the product information in parallel with the product development process. However, it should be noted that this approach only addresses part of the problem. For example, a complex product, such as an aircraft, this issue is still challenging. Thus, future work should look into alternative ways of dealing with this issue.

- Identification of exogenous and constrained parameters

As described in Sect. 6.4, exogenous parameter is defined as the parameter which can be modified directly by the designer and is not dependent on any other parameters. Constrained parameter is defined as the parameter which can only be satisfied through exogenous parameters. In some cases, a parameter may be identified as an exogenous parameter and constrained parameter simultaneously. For example, in the research of Mark and Kosuke [12], the metrics "width", "height", "depth" of a water cooler can be regarded as both exogenous parameters and constrained parameters (customers may make constraints on these metrics). To handle the possible conflict of parameter types, it is assumed that, in the process of constructing product DCMBD models, each parameter can only have one parameter type and the type of exogenous parameter is given priority. In other words, if a parameter is identified as an exogenous parameter, it cannot be regarded as a constrained parameter. In this way, the metrics "width", "height", "depth" in the research of Mark and Kosuke [12] are identified as exogenous parameters. And parameter "size" or "compactness" can be regarded as a constrained parameter, which can be satisfied through the exogenous parameters "width", "height", and "depth".

- Relation between topology faces

One of the assumptions made in this work is that the product is assembled through the constraints between topology faces. However, it is not adopted in all companies to use topology faces to construct the product assembly models. Thus the topology face relations are not always available during the analysis of product design change. One possible way to address this issue is to add the records of topology face constraints in the product DCMBD models. This is to ensure that the improved/changed product can be designed based on the original product through the proposed method in this chapter. As a result, the efficiency of design change with the proposed method can be overestimated. Therefore, more work should be carried out to enhance the applicability.

- Extension to the method

The method presented in this chapter is based on the product DCMBD models. Ideally, the product data in the DCMBD models is sufficient to deal with various changed requirements. However, due to the subjectivity of designers, the data recorded in the product DCMBD models can be incomplete. Hence, there is a need to provide a method to check the data completeness and evaluate its impact on the accuracy of change propagation in the future research.

6.9 Conclusion and Future Work

Using the DCMBD model as the sole data source, this chapter provides a method to analyze the change propagation and allows a more proactive management of engineering changes in product design. Two types of design change are applied in this chapter: parameter change and topology face change, which correspond to two phases of design change. Parameter change and topology face change are mainly aimed at addressing design change impact on the design laws and product assembly, respectively. Four contributions of this chapter are highlighted. First, for efficient product data acquisition, product DCMBD model is regarded as the sole source instead of the product files in different forms. Second, algorithms to acquire the parameter changes and topology face changes are proposed. Third, product hierarchical structure, parameter relation model and topology face relation model are introduced. Fourth, the design change process in a uniform way is proposed, i.e. the parameter design change and topology face design change are alternately applied for the product design change.

In addition to the future work discussed in the previous section, we will test the automatic change propagation process for more products. Further improvements of the prototype system are demanded to generate more comprehensive design change solutions for product. This requires more detailed and accurate data of original product. Besides, a particular cost-benefit analysis of the method will be addressed, and implementation technique needs to be evaluated for efficiency with respect to complex product relations. Although basic in its current form, the MBD-based method has a great potential for development into a powerful design tool. Research to further the work is currently under way.

References

1. Ahmed S, Yethish K (2007) Engineering change during a product's lifecycle. In: International conference on engineering design, ICED'07, Paris, France
2. Fricke E, Gebhard B, Negele H, Igenbergs E (2000) Coping with changes: causes, findings, and strategies. Syst Eng 3(4):169–179
3. Maier AM, Langer S (2011) Engineering change management report 2011: survey results on causes and effects, current practice, problems, and strategies in denmark. DTU Management Engineering, Department of Management Engineering, Technique University of Denmark, Copenhagen
4. Son YK, Savage GJ (2005) Set theoretic formulation of performance reliability of multiple response time-variant systems due to degradations in system components. Qual Reliab Eng Int 23(2):171–188
5. Styblinski MA (1991) Formulation of the drift reliability optimization problem. Microelectron Reliab 31(1):159–171
6. Jorge DC, Gerardo AQ, Pascual C (2013) Enhancing model reuse through 3D annotations: a theoretical proposal for an annotation-centered design intent and design rationale communication. In: Proceedings of the ASME 2013 international mechanical engineering congress and exposition
7. Alemanni M, Destefanis F, Vezzetti E (2011) Model-based definition design in the product lifecycle management scenario. Int J Adv Manuf Technol 52(1–4):1–14
8. Rouibah K, Caskey KR (2003) Change management in concurrent engineering from a parameter perspective. Comput Ind 50:15–34
9. Liu YJ, Lai KL, Dai G et al (2010) A semantic feature model in concurrent engineering. IEEE Trans Autom Sci Eng 7(3):659–665
10. Gao S, Shah JJ (1998) Automatic recognition of interacting machining features based on minimal condition subgraph. Comput Aided Des 30(9):727–739
11. Li YG, Ding YF, Mou WP et al (2010) Feature recognition technology for aircraft structural parts based on a holistic attribute adjacency graph. Proc Inst Mech Eng Part B J Eng Manuf 224(2):271–278
12. Mark VM, Kosuke I (2002) Design for variety: developing standardized and modularized product platform architectures. Res Eng Des 13(2002):213–235

Chapter 7
Matrix-Based Change Prediction and Analysis Method Considering Multiple Change Requirements

7.1 Introduction and Synopsis

This chapter discusses the change propagation in product's structure as a result of multiple change requirements (MCRs). Design projects have been surrounded by tight schedule and cost overruns. Due to undervaluation of design change efforts, project duration has been diagnosed as one of the key reasons for the problem. Firms should resolve engineering changes (ECs) in an economical way with the aim that modified artefacts can be introduced into the market with a competitive cost and time.

First of all, the related literature regarding existing change propagation method is discussed. From the literature, it is revealed that usually single change requirement techniques are used to predict the effect of change propagation in the product. MCRs modeling technique is presented in this chapter for evaluating different change propagation paths (CPPs) in complex product design in perspective of global risk. It outlines the economic consequences of ECs in terms of time, which are made during the product development (PD) process. The design structure matrix (DSM) framework is used to apprehend the dependencies between design tasks/components using the concept of propagation probability and change impact.

The algorithm proposed in this chapter relates the change risk to the amount of rework to be done. It can also be used to compare the total completion time (TCT) of the redesign for the different components sequences. A mathematical model is developed to estimate the TCT of various propagation paths. Finally, the success of the proposed MCR-algorithm is validated through the study of change propagation in the motorcycle engine design process. The result of this analysis is an array of options for a design manager to implement MCRs in a cost-effective way.

Over the last two decades, both academic and industrial concerns in EC propagation have increased. As a result, a variety of methods and tools has been developed by researchers for the prediction and analysis of change propagation. The insight of why and how changes propagate during PD process is crucial

D. Tang et al., *Matrix-Based Product Design and Change Management*,
DOI 10.1007/978-981-10-5077-0_7

because most of the new designs are not produced from the clean sheet, but designed through alterations and changes to the current design [1]. There are always many feasible approaches to design any artefact, and the most efficient firms persistently seek alternative approaches in an effort to enhance their design development process.

In firms that design and produce complex artefacts, specifically custom products, ECs usually arise in the design of the artefact as it evolves. The EC cost is a challenging task to quantify; however, as a design evolves, the cost of change can increase significantly. According to Jarratt et al. [2], the cost of making an engineering change rises by a factor of 10 between each phase of product design and development, and in the same scenario, Evans stated that delay is deadly [3]. The design is almost never free of risk. Therefore, the risk is an indispensable part of engineering design. It exists in each industry and project in a different form and can be handled in a different perspective.

In engineering design processes risk is mostly realized in terms of risk to life and limb because of technical failure or rough handling of the product [4]. Designers try out to quantify risk, however, significance related to these quantities depends upon the design perspective and how they are captured. For a successful completion of a design project, it is extremely significant that the designers should highlight the uncertainties that can pose a possible risk for the project and find out ways of handling such scenario. Risk management is an iterative process and can relate to any aspect of the design project such as; schedule, cost or quality. In order to boost product design processes, managers and designers become more curious that how design tasks should be managed within the process, how rework penetrates through the process, schedule and cost trade-off, and the relationship between all these problems.

Due to the high complexity of engineering products, the change propagation is considered rather probabilistically than deterministically [5]. Most change prediction methods represent the product as a network and describe change propagation as spread within this network because of a single change requirement [6–10]. This chapter suggests a new, MCR-algorithm for change propagation based on the DSM. The problem for design managers is to find out a proper way to implement the MCRs satisfying customer's desires. The repercussion of such a dynamic methodology is that designers must realize and take into consideration interdependencies, relations, and the information that requires being exchanged.

This chapter contributes to the discussion of MCRs, iteration and global risk. Moreover, highlights the challenge of representing it in a process simulation models. The prime aim of this study is to increase the number of considerations that a designer can take into account and therefore help to explore different possible CPPs more thoroughly. The proposed novel technique has the following advantages compared with existing methods as mentioned above.

- Most existing change prediction methods described change propagation as spread within product's network because of a single change requirement. The proposed method can predict the cost-effective CPPs as a result of MCRs in the products' network.

- The suggested MCR modeling technique evaluates different CPPs in complex product design in perspective of global risk. The local risk based existing change prediction techniques do not provide the optimal solutions.
- This method can predict the number of change steps and the design components involved in the change propagation process. Therefore, it can facilitate the designer's understanding of the entire change process.
- The proposed technique can facilitate the designers by providing an optimal sequence of MCR for execution to achieve the cost-effective CPPs.

7.2 Related Literature

ECs are significant motivating factors of product evolution. An efficient and restrained design change is useful to fulfil customer desires, enhance artefact quality, minimize artefact flaws and encourage artefact innovation [6]. If designers can foresee the change risks, scope and path of change propagation before it eventuates and also finds out the best possible solution based on the analysis results than the risk of change can thus be minimized or obviated. In the literature, different techniques are used to model PD process. The classical approaches such as Critical Path Method [11], Program Evaluation and Review Technique [12] are the network based project scheduling methods.

Most of the researchers have added new power to theses classical techniques, which comprises of Monte Carlo sampling [13] to clarify the stochastic task duration, Generalized Precedence Relations [14], Graphical Evaluation and Review Technique [15], that permits feedback loops and probabilistic routing. Iteration is the core characteristic of the design process, and the above-mentioned techniques have limited potential in modeling iterations. Consequently, the researchers turned to the modeling techniques which can deal with the design iterations. Steward [16] was the first to develop DSM to address the issue of component interdependencies and design iteration. Eppinger et al. [17] and Ulrich and Eppinger [18] made extension in work presented by Steward [16] through modeling information links between design components. However, it's hard to explicitly model the sequential interactions between design components with the help of DSM. It is also a challenging task to represent 'Or' and 'And' logic relationship between design components using DSM, when it is enforced in complex product design environment, whereas 'Or/And' logics extensively prevail among design components, in the PD processes [19–21].

These methods have not been adequate if design components durations need to be modified to reflect improvement curve trends or if disruption cost to be evaluated on the basis of MCRs. Effective ways of handling change processes are necessary for attaining high-quality artefacts and efficient development processes. The traditional cost estimating and scheduling techniques focus on assessing costs and schedule after design decisions are made, they do not have the mechanism that allows the designers to perform a rapid 'what if' analysis of the design alternatives and cost analysis.

It is evident from the literature that both, quantitative and qualitative efforts are utilized to foresee, prevent, characterize and control change propagation. These struggles have mostly drawn on network-based analysis by modeling products and change processes as networks of vertices and links. This chapter explains, evaluates and develops models for seeking CPPs from the viewpoint of component linkages. The suggested method combines knowledge from the research field of DSM, ECM and graph theory. The proposed method will help designers to search cost-effective change solutions and will offer guidance about how to tackle the MCRs in an efficient and effective way.

7.2.1 Change Propagation Mechanism in Product Structure

Change propagation is a phenomenon by which one change triggers a sequence of others changes [5]. Simon [22] revealed that the complexity of a product depends upon the relationship between its elements. All engineering products are mostly decomposable because their relationship can never be completely averted. The expression "linkage" was first introduced by Cohen et al. [23], to express the relationship between design tasks and then it was used by Jarratt et al. [24]. Changes can propagate between various design tasks as a result of these linkages.

To examine change propagation in a global perspective, designers must gather the design linkages and assimilate them together. The product models can be categorized into models of parameter linkages [25–27] and models of component linkages [28–31] to incorporate multidisciplinary design linkages and express their change relationships. In this chapter, a model of component linkages is used, and design change propagation is considered as a process in which a preliminary change requirement causes a chain of subsequent changes.

A preliminary change is a change raised by some internal\external source to satisfy the individual's desire and the designer select the component to fulfil a change requirement. A chain of successive changes are the offshoots of the preliminary change. According to Prabhu et al. [32], change propagation can be considered as the cause-effect, cause-effect pattern. It is not necessary that the changes can propagate only to those components, which are directly linked to each other but also to other components, which have indirect connections. Therefore, Beshoy et al. [33] claim that second-order change propagations are most likely to propagate and are difficult to foresee at the time of change. Change propagation affects many aspects of the product design and requires much rework to implement. The objective of change propagation is to amend the design in order to absorb the change requirement and in the meantime make it possible that the design evolves to a new stable condition.

7.2.2 Engineering Change Requirement and Its Role

Change requirement plays an important role in any design process [34, 35]. All firms must do design changes to the artefacts so as to accommodate the change requirements raised in the form of customer requests, legislation requirements, or changing technologies. Change requirements describe participants such as consumers, providers, designers, and manufacturer's needs [36]. Change requirement is expected to arise within any phase of the artefact lifecycle and may cause undesirable complexity and ambiguity within the design process. Most of the expenses involved in managing requirements are a result of change. It has been recognized that requirement change and its management can be expensive and time-consuming [37, 38]. Thus, it is essential to foresee the requirement change propagation as it allows the designers to predict unexpected changes and boost the probability of the project's success. Lots of these changes are properly originated by the customers as new requirements, or by the firms as revised specifications or manufacturing changes. A designer could save time and money if it is possible to make an immediate but accurate evaluation of the overall effects of a design or requirement change before committing to executing a change [31].

Once an artefact has been launched in the market, various change requirements are proposed by the end-users at the start of artefact use, such as: about artefact's working or not meeting the desired needs. A very few change requirements are suggested for most of its lifetime due to stability gained by the product in the market. At the end of product lifecycle, many change requirements are proposed to extend the product life and also to compete with the other similar products in the market. The trend of the number of change requirements over a product's lifetime can be seen in Fig. 7.1.

There are two basic approaches for handling change requirements. Firstly, execute the change requirement on immediate basis due to safety concerns or critical functionality. Secondly, accumulate the change requirements and then implement them at the same time. The latter approach is used to find out the cost-effective CPPs in a complex design process. Different change requirements have different initial risk termed as; planned risk depends on the extent up to which

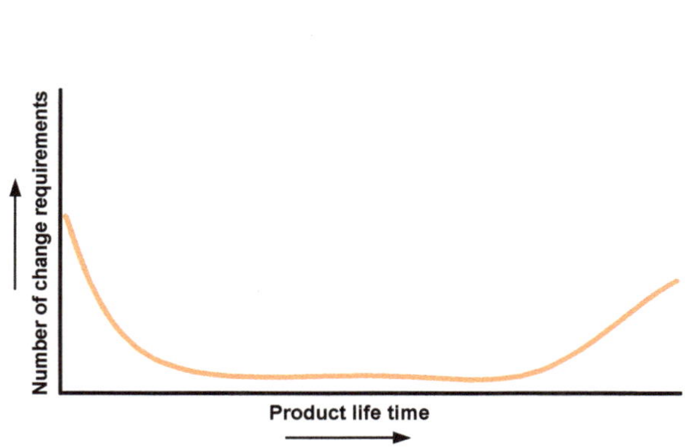

Fig. 7.1 Number of change requirements over product's lifetime (months/years)

the initiating design component should be affected by the change requirement. The foremost objective is to find out the best sequence of the change requirements, in which they should be executed to explore the cost-effective CPPs.

7.2.3 Research Motivation

MCR model presented in this study is applied to incorporate different change requirements raised during PD process that contributes to the change propagation. In this chapter, the cost-effective CPP is described as a path between any two design components in a design process system that can propagate the change to satisfy the MCRs with least cost. The authors examine the following research questions in this study:

- What understandings can be achieved from change propagation of MCRs model?
- How can the suggested model contribute to explore the cost-effective CPP incorporating MCRs?

7.3 Multiple Change Requirements Model

7.3.1 Assumptions

In the proposed MRC-algorithm following four assumptions are made:

- Firstly, when the accumulated risk; propagated change risk and planned risk at any stage of change propagation become greater than 1, then it should be replaced by 1. The reason behind this supposition is that the maximum change in any component is to change the entire design. When the whole component design is changed, it will require 100% of the original design efforts to resolve the design task. Therefore, the accumulated risk cannot exceed the value of 1.
- Secondly, change risk threshold value must be set in order to make the process finite. When the amount of change risk is less than 1% of the initial design work required to resolve the design task, the process of change propagation should be stopped. It is considered that the amount of change risk is too small that it may be neglected.
- Thirdly, in the CPP, when an upstream design component affects the downstream design component to a small amount, the change propagated to the downstream design component will be low and vice versa.
- Fourthly, one change requirement is solely associated with one component.

7.3.2 Description of the Proposed Model

This model assesses the progress of PD process over time. The model can be used to examine various scenarios of the PD process by altering change initiating components and their initial change risk. The model entails information about the different design components that integrate the PD process. The input data are propagation probability, change impact, component design duration and initiating components. The change information flow in the product's structure as a result of MCRs is depicted in Fig. 7.2.

For instance, to demonstrate the concept of change flow between components, consider a product consists of 'n' components as shown in Fig. 7.2. The arrows depict the dependency between components in Fig. 7.2. Several change requirements are raised by different sources to fulfil the desired needs. In Fig. 7.2, 'r_1' to 'r_z' represents different change requirements raised by various sources in a single product and 'C_1' to 'C_n' accounts for the number of product's components. Change requirement initiated in one component has the capability to propagate to the other connected components.

The total implementation cost of desired changes, therefore, comprises of redesigning cost of initial aspects of design and also the other aspects, which are caused due to change propagation. The characterisation of component relationships by their probability and impact allows changes to be considered in terms of risk. Once the component dependencies have been captured, the possible paths of change propagation can be determined. One CPP and the change requirements to the different instigating components are highlighted in green and red colour respectively. With a reference to assumption four, it can be seen in Fig. 7.2, that the number of change requirements 'r_z' must be less than or equal to a number of product's components 'C_n'.

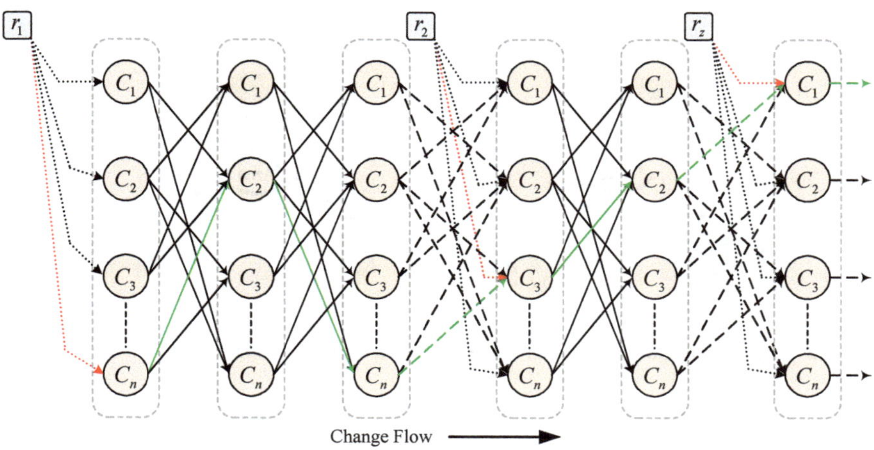

Fig. 7.2 Number of change requirements over product's lifetime

7.3.3 Change Propagation Pattern Based on Logic Relationship Between Components

Products are composed of subsystems, sub-assemblies, and components, which are linked to each other through common interfaces. Due to the interdependencies between components, change propagation takes place within the product. Different logical relationships exist within a product due to components dependencies. Two types of logical relationships, namely, '*OR*' and '*AND*' are considered as shown in Fig. 7.3.

According to logic '*OR*', only one downstream element is influenced by an upstream element at a time. Conversely, in logic '*AND*', multiple downstream elements are affected by an upstream element at once. For instance, if component A_1 propagates changes to only one downstream component (A_2 or A_3), as depicted in Fig. 7.3a, then the relationship between components A_2 and A_3 according to the upstream components A_1 is named as '*OR*'. Contrariwise, if component A_1 spreads changes to downstream components A_2 and A_3 at the same time, as presented in Fig. 7.3b, then the relationship between components A_2 and A_3 according to the upstream components A_1 is labelled as '*AND*'. Due to the presence of logical relationships between elements, two types of change propagation pattern exist in the design process, i.e., parallel and sequential change propagation as shown in Fig. 7.4a, b.

In the case of parallel change propagation, downstream elements originating from the same upstream element are handled simultaneously, if they are not dependent on each other as shown in Fig. 7.4a. In sequential change propagation, each element is in a queue and is treated one after another as illustrated in Fig. 7.4b. From Fig. 7.4a it is evident that parallel change propagation pattern comprises of multiple sequential CPPs. Numerous CPPs are available in the design process due to the presence of logical relationships.

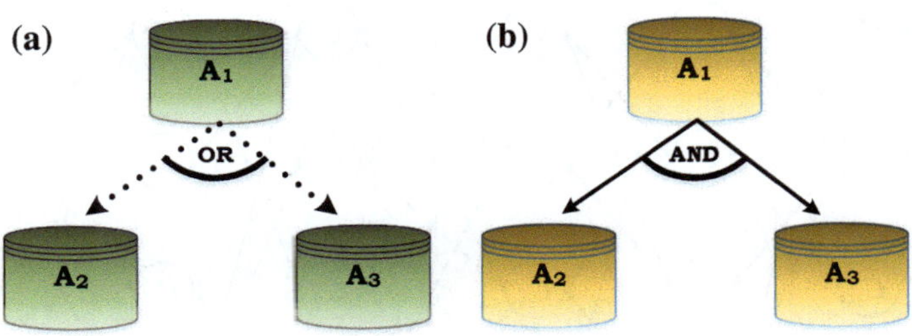

Fig. 7.3 Logical relationship between the components **a** OR, **b** AND

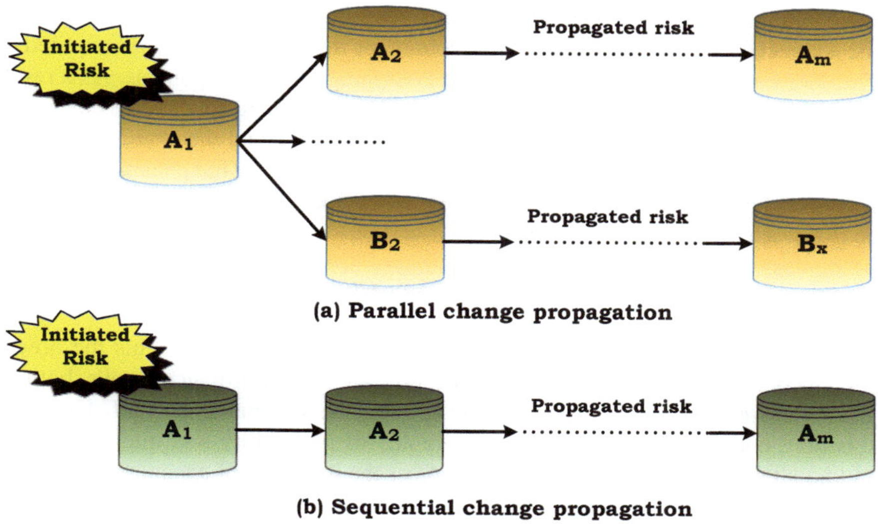

Fig. 7.4 Parallel and sequential change propagation pattern

7.4 Change Risk Propagation and Assessment

Propagation paths are the chain of changes between the initiating and the end component. These chains of changes consist of two different types of changes risk, termed as planned risk and propagated risk.

7.4.1 Change Propagation Mathematical Model for Multiple Change Requirements

Requirement change happens over the entire PD process. Changes can propagate into the product's structure using two variables, i.e., propagation probability and change impact. Several CPPs can be achieved by taking information flow into consideration between two directly connected components. Propagation probability 'PP_{ij}' can be defined as "the average chance that change initiated in one part will cause a change to another connected part through their mutual interface". However, change impact 'CI_{ij}' is the influence of one part over the other linked part as a result of changes and can be defined as "the work needed to redesign a part due to propagated changes".

The change propagation between the linked components is owing to probability and impact, permit changes to be handled in terms of risk. Risk is expressed as "it is the probability of occurring of an activity times the severity of the effects" [39]. Quantitative risk-based analysis technique is proposed to explore the distinct CPPs.

Information flow between activities usually precedes to rework in projects [40]. In the advanced model, propagated risk accounts for the amount of rework required to be done. For instance, consider a product consisting of 'n' components. As a result of change initiation, component 'C_i' passes on the change to directly connected component 'C_j' with the propagated probability of 'PP_{ij}' and the change impact of 'CI_{ij}'. Then, according to Van Bossuyt et al. [39] and Zhang et al. [40], the propagated risk 'PR_{ij}' between two directly connected components 'C_i' and 'C_j' can be given as:

$$PR_{ij} = PP_{ij} \times CI_{ij} \tag{7.1}$$

In Eq. (7.1), 'i' denotes the upstream component and 'j' represents the downstream component in the product's architecture. In DSM, for diagonal cells, both 'i' and 'j' are equal, which means that it denotes the internal impact of the corresponding component. In such cases, the value of probability and impact are zero and no attention is being given in DSM. After the initiation of change requirement, the designer explores the originating component, which must be altered to fulfill the need. This modification is compulsory, and it is termed as initiated change. In this chapter, 'PP_{ri}' and 'CI_{ri}' represent the propagation probability and change impact of initiated change respectively and instigate the risk to the initiated component as a result of change requirement. This change requirement has some impact on the component in terms of redesign cost. This type of risk is known as planned risk because it must be implemented to accomplish the change requirement, and can be expressed as:

$$P_lR(r, c_i) = PP_{r,i} \times CI_{r,i} \tag{7.2}$$

where in Eq. (7.2), $P_lR\ (r,\ c_i)$ is the planned risk, due to the change requirement 'r' to be implemented in the initiating component 'C_i'. $PP_{r,i}$ and $CI_{r,i}$ are the propagation probability and the change impact, which causes initiated changes because of the change requirement 'r' respectively. Once the change requirement is implemented, designers further find out that the changes propagate to further associated components and will also amend them. The change propagation between the components is due to propagated risk, which is the outcome of the planned risk. The propagated risk between two components can be expressed as:

$$P_rR(c_i, c_1) = PP_{i,1} \times CI_{i,1} \tag{7.3}$$

where in Eq. (7.3), $P_rR\ (c_i,\ c_1)$ is the propagated risk between the initiating component 'C_i' and the directly connected component 'C_1'. Similarly, change propagation takes place throughout the entire product architecture. Different CPPs are available in the product's architecture to implement the change requirements. The possibility to evaluate all the propagation paths between two components can explore valuable information regarding the whole product. The propagation probability and the change impact are defined for those components, which are directly connected to each other. However, the effect of change can propagate to all other components in the product's structure, which are connected directly or indirectly. By considering both effects of change propagation, the predictive model must be extended to calculate the cumulative change risk. The cumulative change propagated risk contains both, planned and propagated risk.

$$\begin{aligned} CPR^{k}_{r(ij)} &= P_l R(r, c_i) \times P_r R(c_i, c_1) \times P_r R(c_1, c_2) \times \cdots \times P_r R(c_{k-2}, c_{k-1}) \\ &\quad \times P_r R(c_{k-1}, c_j) \end{aligned} \tag{7.4}$$

where in Eq. (7.4), $CPR^{k}_{r(ij)}$ is the change risk propagated between the initiating component 'C_i' and the last component 'C_j' in the CPP to accomplish the change request 'r' and 'k' is the number of change steps between the initiating component and the last component. The change propagates from initiating component 'C_i' to the last component 'C_j' after 'k' transmissions through '$k-1$' components. If two change requirements are desired by the customer in a one product, then these alterations must be incorporated at once to minimize the redesign cost, and the distraction in production and the Eq. (7.4) can be modified as:

$$\left.\begin{aligned} CPR^{K}_{r_1 r_2(ij)} &= [\{P_l R(r_1, c_{i_1}) \times P_r R(c_{i_1}, c_1^{r_1}) \times P_r R(c_1^{r_1}, c_2^{r_1}) \times \cdots \times P_r R(c_{k_1-2}^{r_1}, c_{k_1-1}^{r_1}) \\ &\quad \times P_r R(c_{k_1-1}^{r_1}, c_{i_2})\} + P_l R(r_2, c_{i_2})] \times P_r R(c_{i_2}, c_1^{r_2}) \times P_r R(c_1^{r_2}, c_2^{r_2}) \times \cdots \\ &\quad \times P_r R(c_{k_2-2}^{r_2}, c_{k_2-1}^{r_2}) \times P_r R(c_{k_2-1}^{r_2}, c_j) \end{aligned}\right\} \tag{7.5}$$

Similarly, for 'z' change requirements Eq. (7.5) can be expressed as:

$$\left.\begin{aligned} CPR^{K}_{r_1 r_2 \cdots r_z(ij)} &= [\cdots[[P_l R(r_1, c_{i_1}) \times P_r R(c_{i_1}, c_1^{r_1}) \times \cdots \times P_r R(c_{k_1-2}^{r_1}, c_{k_1-1}^{r_1}) \times P_r R(c_{k_1-1}^{r_1}, c_{i_2})] \\ &\quad + P_l R(r_2, c_{i_2})] \times P_r R(c_{i_2}, c_1^{r_2}) \times \cdots \times P_r R(c_{k_2-2}^{r_2}, c_{k_2-1}^{r_2}) \times P_r R(c_{k_2-1}^{r_2}, c_{i_3})] \\ &\qquad \vdots \\ &\qquad \vdots \\ &\quad + P_l R(r_z, c_{i_z})] \times P_r R(c_{i_z}, c_1^{r_z}) \times \cdots \times P_r R(c_{k_z-2}^{r_z}, c_{k_z-1}^{r_z}) \times P_r R(c_{k_z-1}^{r_z}, c_j)] \end{aligned}\right\} \tag{7.6}$$

where in Eq. (7.6), $CPR^{K}_{r_1 r_2 \cdots r_z(ij)}$ is the cumulative change risk propagated between initiating component 'C_i' and the last component 'C_j' in the CPP to accomplish 'z' number of change requests 'r'. 'C_{iz}' represents the initiating components for a particular change requirement. $C^{r_z}_{k_z-2}$ accounts for the intermediate components for a particular change requirement in the CPP. If 'z' is the number of change requirements and 'n' is the number of product's component then $z \leq n$. 'K' represents the entire set of change step in the CPP, initiating from components 'C_i' and propagates to the last component 'C_j', to accomplish the desired 'z' change requirements in a single product. 'k_1' up to 'k_z' represents the number of change steps in the same propagation path for a particular change requirement as shown in Eq. (7.7).

$$K = k_1 + k_2 + k_3 + \cdots + k_z \tag{7.7}$$

In a product design, the iteration of earlier design activity is common. Iterations are due to the interdependent design components, in which the information is required from each other. The approximation of product development cost and time is difficult because of iterations, complex interdependencies and change requirements. Design iteration between components usually occurs as a result of incomplete information [41]. Each iteration helps to improve the results and enhance the product quality. Designer's knowledge regarding product design increases as a result of each iteration and therefore less effort is required to complete the design [42]. In the suggested mathematical model, the estimation of development duration is facilitated by including the learning effect. During the change propagation process, the component affected by an upstream component to a small amount will also influence the downstream component to a slight extent and vice versa. Therefore, change risk propagated to the downstream component depends on the change risk of an upstream component. Thus, the risk propagated from component 'C_i' to 'C_j' in the 'Ith' change iteration, considering the learning effect, can be given as follows:

$$PR_{ij}(I) = (PP_{ij} \times CI_{ij}^{I}) \times PR_{i-1,i} \tag{7.8}$$

To build an accurate model, Andersson et al. described a curve termed as "learning-by-doing", in which the execution time decreases due to each design iteration with an associated learning curve function [43]. Therefore, Eq. (7.8) is derived based on the assumption that the component change impact has a nonlinear relationship with the number of iterations during change propagation process between each pair of components, to incorporate the learning effect. The 'Ith' change iteration from component 'C_i' to 'C_j' is the learning effect developed by the designers as a result of component design familiarity. In Eq. (7.8), '$PR_{i-1,i}$' is the risk propagated from component 'C_{i-1}' to 'C_i'. The TCT to resolve the engineering changes as a result of MCRs can be computed as below:

$$TCT = \sum_{i=1}^{m} CPR_{i-1,i} \times CT_{c_i} \tag{7.9}$$

In Eq. (7.9), 'm' represents the number of changed components involved in the CPP as a result of a change requests received. 'CT_{c_i}' is the original completion time of component 'C_i' in days and ' $CPR_{(i-1,i)}$ is the amount of change propagated risk from an upstream element to downstream component 'C_i'. The value of change propagated risk will be in the range of 0–1. Therefore, redesign duration will always be less than the initial design efforts. The risk propagated to each component is different. Each CPP consists of numerous components. As a result, the TCT of propagated path is the sum of redesign time of each changed element.

7.4.2 Illustration of the Proposed Mathematical Model

For instance, consider a simple product that consists of four components to illustrate the idea presented in the above section. To acquire the inter-dependencies between the product's components, propagation probability (PP_{ij}) and change impact (CI_{ij}) matrices are taken into consideration as shown in Figs. 7.5 and 7.6 respectively. The entries in the off-diagonal cells point out the direct relationship between the components, which are directly connected to each other. The completion time of each component is presented in Table 7.1.

Let us suppose that two change requests are raised by the customer at the same time. Now the designer determines the initiating components to meet the change requirement. Designer pointed out that 'C_2' and 'C_4' are the components, which must be changed to fulfill the customer requirement. In this scenario, the number of

Fig. 7.5 Probability matrix of a small change propagation model

Affected Components (columns); *Initiated Components* (rows)

	C_1	C_2	C_3	C_4
C_1		0.4		0.3
C_2			0.5	0.8
C_3	0.4			
C_4		0.8	0.6	

$PP= [PP_{ij}]_{4x4}$

Fig. 7.6 Impact matrix of a small change propagation model

Affected Components (columns); *Initiated Components* (rows)

	C_1	C_2	C_3	C_4
C_1		0.8		0.5
C_2			0.3	0.7
C_3	0.6			
C_4		0.3	0.9	

$CI= [CI_{ij}]_{4x4}$

Table 7.1 Design duration of different components

S. no.	Design components	Design duration (days)
1	C_1	10
2	C_2	7
3	C_3	14
4	C_4	4

change requirement 'z' is 2 and the number of product's component 'n' is 4, which satisfies the condition:

$$z \leq n \tag{7.10}$$

The initial change to the initiating component due to change requirement is termed as planned risk ($P_l R$). The change will further disseminate to the other connected components throughout the product structure, and it is termed as propagated risk ($P_r R$). Let's assume that components 'C_2' and 'C_4' have the initial change impact of 0.5 and 0.7 respectively. According to Eq. (7.2) planned risk for both the components can be calculated as follows:

$$\begin{aligned}
&\textit{For component } C_4\\
&PP_{r_1,i} = 1\\
&CI_{r_1,i} = 0.7
\end{aligned}$$

$$\begin{aligned}
&\textit{For component } C_2\\
&PP_{r_2,i} = 1\\
&CI_{r_2,i} = 0.5
\end{aligned}$$

Here 'i' represents the instigating component with respect to corresponding change requirement 'r'. Therefore, 'i' is replaced by 4 and 2 in both cases respectively. First of all, we will calculate the planned risk for both the components.

$$\begin{aligned}
&\textit{Planned risk for component } 'C_4' \textit{ due to change requirement } 'r_1'\\
&P_l R_{r_1}(r,4) = PP_{r_{1,4}} \times CI_{r_{1,4}}\\
&P_l R_{r_1}(r,4) = 1 \times 0.7\\
&P_l R_{r_1}(r,4) = 0.7
\end{aligned}$$

$$\begin{aligned}
&\textit{Planned risk for component } 'C_2' \textit{ due to change requirement } 'r_2'\\
&P_l R_{r_2}(r,2) = PP_{r_{2,2}} \times CI_{r_{2,2}}\\
&P_l R_{r_2}(r,2) = 1 \times 0.5\\
&P_l R_{r_2}(r,2) = 0.5
\end{aligned}$$

In the illustrated example the designer decided to start the change propagation process from component 'C_4'. It can be seen from Figs. 7.5 and 7.6 that component 'C_4' has a direct relation with the components 'C_2' and 'C_3'. For simplicity, only the change to one component 'C_3' is considered. The propagation probability (PP_{43}) and change impact (CI_{43}) between components 'C_4' and 'C_3' are 0.6 and 0.9 respectively as shown in Figs. 7.5 and 7.6. Thus, the change risk that propagates between components 'C_4' and 'C_3' can be calculated using Eq. (7.3).

$$P_rR_{r_1}(4,3) = PP_{4,3} \times CI_{4,3}$$
$$P_rR_{r_1}(4,3) = 0.6 \times 0.9$$
$$P_rR_{r_1}(4,3) = 0.54$$

The overall risk propagated to component 'C_3' as a result of change request 'r_1' can be given by using Eq. (7.4).

$$P_lR_{r_1}(r,4) = 0.7$$
$$P_rR_{r_1}(4,3) = 0.54$$
$$CPR_{r_1(r,3)} = P_lR_{r_1}(r_1,4) \times P_rR_{r_1}(4,3)$$
$$CPR_{r_1(r,3)} = 0.7 \times 0.54$$
$$CPR_{r_1(r,3)} = 0.378$$

The changes will further propagate to the other associated components and can be calculated as given below:

$$P_rR_{r_1}(3,1) = PP_{3,1} \times CI_{3,1}$$
$$P_rR_{r_1}(3,1) = 0.4 \times 0.6$$
$$P_rR_{r_1}(3,1) = 0.24$$

$$CPR_{r_1(r,1)} = P_lR_{r_1}(r_1,4) \times P_rR_{r_1}(4,3) \times P_rR_{r_1}(3,1)$$
$$CPR_{r_1(r,1)} = 0.378 \times 0.24$$
$$CPR_{r_1(r,1)} = 0.091$$

$$P_rR_{r_1}(1,4) = PP_{1,4} \times CI_{1,4}$$
$$P_rR_{r_1}(1,4) = 0.3 \times 0.5$$
$$P_rR_{r_1}(1,4) = 0.15$$

$$CPR_{r_1(r,4)} = P_lR_{r_1}(r_1,4) \times P_rR_{r_1}(4,3) \times P_rR_{r_1}(3,1) \times P_rR_{r_1}(1,4)$$
$$CPR_{r_1(r,4)} = 0.091 \times 0.15$$
$$CPR_{r_1(r,4)} = 0.0136$$

$$P_rR_{r_1}(4,2) = PP_{4,2} \times CI_{4,2}$$
$$P_rR_{r_1}(4,2) = 0.8 \times 0.3$$
$$P_rR_{r_1}(4,2) = 0.24$$

$$CPR_{r_1(r,2)} = P_lR_{r_1}(r_1,4) \times P_rR_{r_1}(4,3) \times P_rR_{r_1}(3,1) \times P_rR_{r_1}(1,4) \times P_rR_{r_1}(4,2)$$
$$CPR_{r_1(r,2)} = 0.0136 \times 0.24$$
$$CPR_{r_1(r,2)} = 0.0032$$

When the change risk propagated from initial component to some intermediate component in the product's structure is less than 1% of the initial design work required to resolve the design activity, the propagation should be stopped. It is used as a termination criterion to finish the process, as the effect of such change risk is too small that it can be neglected. Different threshold values can be used as a stop criterion to terminate the propagation process depending on the design project.

In the above calculation the change risk propagated from 'C_4' to 'C_2', as a result of change requirement 'r_1' is less than 1%. The propagation should be terminated at this point, but at the same time 'C_2' is the initiating component for change requirement 'r_2'. Therefore, the total amount of change risk transferred to 'C_2' is the sum of change propagated risk $CPR_{r_1(r,2)}$ and the planned risk $P_lR_{r_2}$ $(r, 2)$. As a result, the propagation process will not terminate at this point, as the sum of the change risk is greater than the threshold value of the stop criterion. The cumulative risk is calculated by using Eq. (7.5).

$$\textit{The planned risk for component } 'C_2'$$
$$P_lR_{r_2}(r,2) = PP_{r_2,2} \times CI_{r_2,2}$$
$$P_lR_{r_2}(r,2) = 1 \times 0.5$$
$$P_lR_{r_2}(r,2) = 0.5$$

$$CPR_{r_1r_2(r,2)} = [P_lR_{r_1}(r_1,4) \times P_rR_{r_1}(4,3) \times P_rR_{r_1}(3,1) \times P_rR_{r_1}(1,4) \times P_rR_{r_1}(4,2)] + P_lR_{r_2}(r,2)$$
$$CPR_{r_1r_2(r,2)} = 0.0032 + 0.5$$
$$CPR_{r_1r_2(r,2)} = 0.5032$$

The process will continue to propagate the changes to the associated components until it reaches the threshold value.

$$\begin{aligned}P_rR_{r_2}(2,4) &= PP_{2,4} \times CI_{2,4}\\ P_rR_{r_2}(2,4) &= 0.8 \times 0.7\\ P_rR_{r_2}(2,4) &= 0.56\end{aligned}$$

$$\begin{aligned}CPR_{r_1r_2(r,4)} &= [[P_lR_{r_1}(r_1,4) \times P_rR_{r_1}(4,3) \times P_rR_{r_1}(3,1) \times P_rR_{r_1}(1,4) \times P_rR_{r_1}(4,2)]\\ &\quad + P_lR_{r_2}(r,2)] \times P_rR_{r_2}(2,4)\\ CPR_{r_1r_2(r,4)} &= 0.5032 \times 0.56\\ CPR_{r_1r_2(r,4)} &= 0.2818\end{aligned}$$

$$\begin{aligned}P_rR_{r_2}(4,2)(2)^* &= PP_{4,2} \times CI_{4,2}\\ P_rR_{r_2}(4,2)(2)^* &= 0.8 \times 0.3^{2^*}\\ P_rR_{r_2}(4,2)(2)^* &= 0.072\end{aligned}$$

The numeric values, highlighted with an asterisk represent the number of design iterations. The change risk propagated between two components depends on how many design iterations have been performed. Change risk is a declining function depending upon the number of iterations executed because the value of change impact is between 0 and 1. In each iteration, the design information refines, and thus the designer takes lesser time to accomplish the design task.

$$\begin{aligned}CPR_{r_1r_2(r,2)} &= [[P_lR_{r_1}(r_1,4) \times P_rR_{r_1}(4,3) \times P_rR_{r_1}(3,1) \times P_rR_{r_1}(1,4) \times P_rR_{r_1}(4,2)]\\ &\quad + P_lR_{r_2}(r,2)] \times P_rR_{r_2}(2,4) \times P_rR_{r_2}(4,2)(2)\\ CPR_{r_1r_2(r,2)} &= 0.2818 \times 0.072\\ CPR_{r_1r_2(r,2)} &= 0.0209\end{aligned}$$

$$\begin{aligned}P_rR_{r_2}(2,4)(2)^* &= PP_{2,4} \times CI_{2,4}^2\\ P_rR_{r_2}(2,4)(2)^* &= 0.8 \times 0.7^{2^*}\\ P_rR_{r_2}(2,4)(2)^* &= 0.392\end{aligned}$$

$$\begin{aligned}CPR_{r_1r_2(r,4)} &= [[P_lR_{r_1}(r_1,4) \times P_rR_{r_1}(4,3) \times P_rR_{r_1}(3,1) \times P_rR_{r_1}(1,4) \times P_rR_{r_1}(4,2)]\\ &\quad + P_lR_{r_2}(r,2)] \times P_rR_{r_2}(2,4) \times P_rR_{r_2}(4,2)(2) \times P_rR_{r_2}(2,4)(2)\\ CPR_{r_1r_2(r,4)} &= 0.0209 \times 0.392\\ CPR_{r_1r_2(r,4)} &= 0.0082\end{aligned}$$

The change propagation process will stop at this point because the propagated risk is less than the threshold value. The CPP achieved as a result of change requests 'r_1' and 'r_2' can be given as follows:

$$\textit{Change propagation path} \quad C_4^* \rightarrow C_3 \rightarrow C_1 \rightarrow C_4 \rightarrow C_2^* \rightarrow C_4 \rightarrow C_2 \rightarrow C_4$$

In the above CPP, component 'C_4' is the instigating component 'i' and also the ending component 'j'. When the component 'C_2' arrives the first time in the CPP, it will also act as an instigating component for change requirement 'r_2'. The number of change steps 'k' in the CPP is 7, and the intermediate components '$k - 1$' between the instigating and ending components are 6 through which the change propagation take place.

$$\textit{Change propagation path} \quad C_4^* \rightarrow C_3 \rightarrow C_1 \rightarrow C_4 \rightarrow C_2^* \rightarrow C_4 \rightarrow C_2 \rightarrow C_4$$

In the above CPP the information flow from component 'C_4' to 'C_2' and from component 'C_2' to 'C_4' take place twice. Therefore, the learning effect is applied in the illustrative example to minimize the redesign duration because of the designer's familiarity with the component's design. The TCT for resolving the change requirements 'r_1' and 'r_2' can be computed by using Eq. (7.9).

$$TCT = \sum_{i=1}^{m} CPR_{i-1,i} \times CT_{C_i}$$

Where 'm' represents the number of changed components as a result of change requests 'r_1' and 'r_2' in a single CPP. The value of 'm' in the illustrated example is 8 as shown in the above propagation path. Therefore, replace 'm' by 8 in the above equation and then expand it. The following result is achieved.

$$TCT = \sum_{i=1}^{m} CPR_{i-1,i} \times CT_{C_i}$$

Put $m = 8$ *in the above equation*

$$TCT = \sum_{i=1}^{8} CPR_{i-1,i} \times CT_{C_i}$$

After Expension

$$\begin{aligned} TCT = {} & CPR_{1-1,1} \times CT_{C_1} + CPR_{2-1,2} \times CT_{C_2} + CPR_{3-1,3} \times CT_{C_3} + CPR_{4-1,4} \times CT_{C_4} \\ & + CPR_{5-1,5} \times CT_{C_5} + CPR_{6-1,6} \times CT_{C_6} + CPR_{7-1,7} \times CT_{C_7} + CPR_{8-1,8} \times CT_{C_8} \end{aligned}$$

$$\begin{aligned} TCT = {} & CPR_{0,1} \times CT_{C_1} + CPR_{1,2} \times CT_{C_2} + CPR_{2,3} \times CT_{C_3} + CPR_{3,4} \times CT_{C_4} \\ & + CPR_{4,5} \times CT_{C_5} + CPR_{5,6} \times CT_{C_6} + CPR_{6,7} \times CT_{C_7} + CPR_{7,8} \times CT_{C_8} \end{aligned}$$

In the above equation component, 'C_1' to 'C_8' are those elements, which take part in the change propagation.

Change propagation path	4 →	3 →	1 →	4 →	2 →	4 →	2 →	4
	↓	↓	↓	↓	↓	↓	↓	↓
Components description	C_1	C_2	C_3	C_4	C_5	C_6	C_7	C_8

The original design completion time of each component is given in Table 7.1. Put the values in the above equation to get the TCT of the propagated path.

$$\left.\begin{array}{l} TCT = 0.7 \times 4 + 0.378 \times 14 + 0.091 \times 10 + 0.0136 \times 4 + 0.5032 \times 7 + 0.2818 \times 4 \\ \quad + 0.0209 \times 7 + 0.0082 \times 4 \end{array}\right\}$$

$$\begin{aligned} TCT &= 2.8 + 5.292 + 0.91 + 0.0544 + 3.5224 + 1.1272 + 0.1463 \\ &\quad + 0.0082 \times 4 \\ TCT &= 13.85\ days \end{aligned}$$

7.5 Cost-Effective CPPs Taking MCRs into Consideration

To find the most economical CPPs, the aforementioned mathematical model and logical relationships between the design components are considered. An advanced MCR-algorithm is suggested to generate the entire set of possible propagation paths with the TCT. In the proposed algorithm, MCRs caused by different sources are handled simultaneously to find out the most time-saving CPPs. The MCR-algorithm also considers the iteration phenomenon between the design components. The process flow chart of the MCR-algorithm is showed in Fig. 7.7.

CPP can be defined as "a finite sequence of elements and the direct dependencies between them". Due to the presence of cyclic interdependencies between components, parameters and system's interfaces, advancement in the design cannot be accomplished through a linear process. Matrix-based approaches are capable of handling loops and cycles. Propagation impacts have been widely studied in change prediction models based on product structure DSMs. However, in these models, design change completion time and cost were not considered. The easiest approach for the task rework duration is to presume fix, deterministic task duration that does not alter on repeated efforts [44]. However, it may be very simplistic approach [45]. From the literature, it is revealed that the effort required to redesign an element is less than the effort to design it for the first time. The proposed algorithm also considers the reduction in component duration due to consecutive iterations.

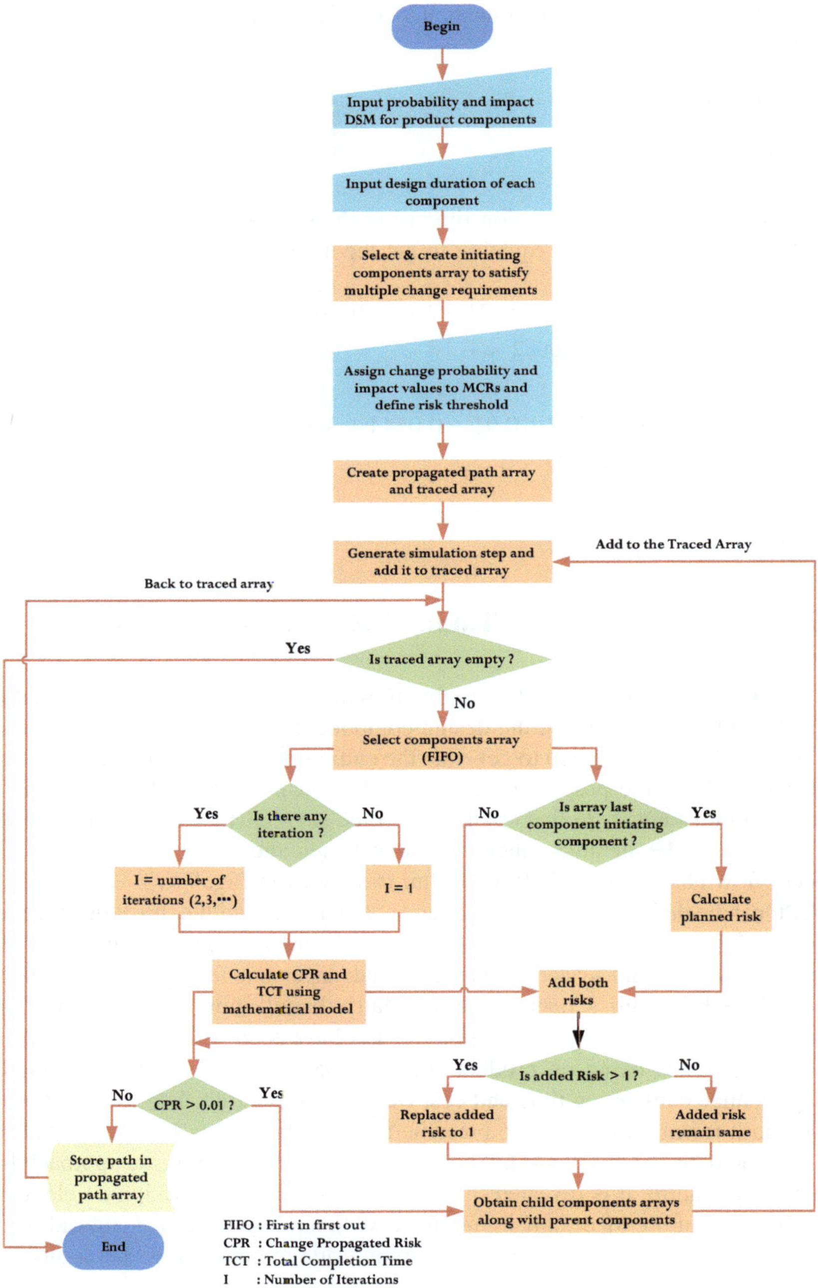

Fig. 7.7 Process flow chart for MCRs

The proposed algorithm is obtained by doing some alterations in the Breadth first search (BFS) technique, which is used to find the shortest possible route between the reachable nodes. BFS technique can traverse a graph or tree data structure. The graph-based search method is used to incorporate the iteration phenomenon because graph contains cycles. In this approach, the neighboring nodes are first explored before moving to the other nodes. Queue data structure is applied in the algorithm, which works on the rule of first-in, first-out (FIFO). The components are taken away from the queue in the same order in which they were entered.

In the suggested MCR-algorithm three data arrays are presented, named as, traced array, propagated path array and initiating component array. Traced array is used to keep track of the in-process CPPs, while propagated path array keeps the record of the finished propagated paths. Initiating component array stores the data of change originating components that depend on the number of change requirements raised. The in-process CPPs are stored in the form of layers in traced array, and it serves on the basis of FIFO queue.

First of all, input the values of propagation probability and change impact between the product's components in the form of DSM. Design component duration is also stored in the form of an array. The designer must investigate the initiating components, which satisfy the change requirements raised by the customer and store that record in the initiating component array. When the simulation is run, the planned risk from the change requirement 'r_1' is calculated and all the child design tasks along with the parent design task are stored in the succeeding layers of the traced array.

In each step, the process will check the repetition of the design component pair to add the learning effect in the process, and at the same time, it will also check that the new design component in an initiating component or not. In the case of an initiating component, the planned risk due to change requirement 'r_z' will be calculated, and then it will be added to the propagated risk. If the accumulated risk is greater than 1, it will be replaced with 1, which means that complete design of the component will be changed.

After that the change risk is checked, if it is less than the defined threshold value (1% of the initial design effort), the path will be stored in the propagated path array along with their completion time. The process will go back to the traced array to pick the subsequent path and repeat the process. On the other hand, if change risk value is greater than the defined threshold value, it will pick all the child design components and add it to the traced array along with the parent design task in the succeeding layer and then repeat the process. The simulation will be executed until the traced array is empty. If the entire accomplishment interval of the succeeding propagation path is less than the earlier propagation path, then it will be stored in the propagated path array earlier than the previous path. Thus, all CPPs from different initiating components to other associated components can be explored.

7.6 Application of MRC-Algorithm

In this section, the author elaborated that how the proposed methodology will assist the designers in re-designing the project to implement the MCRs. To demonstrate the practical utility of suggested model, the author discusses a case study of motorcycle engine design. The process model is built by referring to the engine component-based DSM model comprising of 26 basic design components in Tang et al. [46]. The component-based DSM model consists of two matrices, i.e., propagation probability matrix and change impact matrix, as shown in Figs. 7.8 and 7.9 respectively. Both the numeric DSMs illustrate the network of components direct dependency and provide information regarding propagation probability and change impact between engine elements. The digit in any off-diagonal cell represents the direct dependency between the corresponding components. The dependency includes component interface information and other information resources transferred between design components.

For each element, its column shows its inflow information and its row displays its outflow information. Some DSM uses the different convention; both formats deliver the same information. If the activities are itemized in DSM, then the digits above the diagonal represent the feed forward information from upstream tasks to the downstream components in the process model. Conversely, the digits below the diagonal denote the feedback information from downstream components to the upstream components in the process model. The feedback flow of information is because of iterations in the design process. The additional information regarding the design duration of each component is taken from the article of Li et al. [47]. The unit of design duration is 'day', and the values are shown in Table 7.2.

For instance, if a design change occurs in cylinder head gasket (component number 4) as a result of the change request in the engine design process. Then according to Fig. 7.8, cylinder head gasket has a 60% chance of causing change propagation to the cylinder head (Component number 2) with a change impact of 50% as shown in Fig. 7.9. As a result, the risk propagated from cylinder head gasket to cylinder head can be determined by multiplying both values, i.e., ($0.6 \times 0.5 = 0.3$). The lower and upper limits of both, propagation probability and change impact are between 0 and 1 respectively. In both DSMs, if the numeric values in the off-diagonal cells are in between 0 and 1, then it represents the probability and impact of propagated changes. If the value is 0 or the cell is empty, then it denotes that no changes will propagate to the other components, which means that components are not dependent on each other. The same can be described as below:

$$PP_{ij}/CI_{ij} = \left\{ \begin{array}{lll} empty & (i = j) & Represents\ reflexive\ relation \\ 0 < PP_{ij}/CI_{ij} > 1 & (i \neq j) & Represents\ propagated\ changes \\ 0\ or\ empty & (i \neq j) & Represents\ no\ dependency \end{array} \right\}$$

Design Components		1	2	3	4	5	6	7	8	9	10	11	12	13	14	15	16	17	18	19	20	21	22	23	24	25	26
Air Valve	**1**		.2									.6							.5		.3		.5			.3	
Cylinder Head	**2**				.6														.8								
Crankshaft Pin	**3**								.6											.6	.3						
Cylinder Head Gasket	**4**		.6																.4								
Crankcase	**5**												.3	.2				.3	.3	.4	.2	.4			.3	.2	.7
Piston	**6**										.8			.6					.2		.3			.8			
Gaseous Ring	**7**						.6												.3					.8			
Rear Crankshaft	**8**			.6																.8	.3	.7					
Oil Ring	**9**						.7												.1					.6			
Piston Pin	**10**						.8							.4							.3						
Valve Adjustable Screw	**11**	.6																					.8				
Strainer	**12**					.4												.5									.5
Connecting Rod	**13**					.3	.6				.4					.8	.4			.8							
Camshaft	**14**					.3														.6	.3					.8	
Connecting Rod Journal	**15**													.8			.5			.7	.3						
Connecting Rod Bearing	**16**					.3								.4		.5					.3						
Engine Oil Pump	**17**					.3							.5								.3						.6
Cylinder Block	**18**	.3	.7		.8	.4	.2																	.6			
Front Crankshaft	**19**			.6		.8			.8					.8	.6	.7											
Oil Passage	**20**	.5		.5		.4	.5		.5		.5				.5	.5	.5	.8		.5				.5			.4
Flywheel	**21**								.7																		
Valve Retainer Nut	**22**	.3										.3													.4		
Cylinder Liner	**23**											.8							.6		.3						
Valve Guide	**24**						.2			.4																	
Cam	**25**	.5				.4																			.2		
Oil Pan	**26**					.7												.6			.4						

Fig. 7.8 DSM showing propagation probabilities of engine parts

A number of change requirements can be raised by numerous sources (internal/external). To fulfill these change requirements, initially, the designers must investigate to find the initiating components. For instance, if in the motorcycle engine design process two ECs are raised (internal/external), and the designer investigates that it causes to change the components "Piston" and "Cylinder Block" design. The next step for the designer is to explore the best sequence of the

Design Components		1	2	3	4	5	6	7	8	9	10	11	12	13	14	15	16	17	18	19	20	21	22	23	24	25	26
Air Valve	1		.2									.2							.7		.3		.4			.6	
Cylinder Head	2				.7														.9								
Crankshaft Pin	3								.4											.3	.2						
Cylinder Head Gasket	4		.5																.2								
Crankcase	5												.8	.6				.7	.7	.5	.4	.2			.4	.8	.7
Piston	6										.2			.6					.3		.4			.8			
Gaseous Ring	7						.8												.8					.6			
Rear Crankshaft	8			.5																.8	.3	.2					
Oil Ring	9						.2												.2					.6			
Piston Pin	10						.6							.4							.3						
Valve Adjustable Screw	11	.2																					.8				
Strainer	12					.4												.5									.3
Connecting Rod	13					.7	.5				.9					.8	.4			.8							
Camshaft	14					.3														.3	.6					.8	
Connecting Rod Journal	15													.6			.6			.6	.4						
Connecting Rod Bearing	16					.5								.7		.4					.7						
Engine Oil Pump	17					.5							.6								.8						.6
Cylinder Block	18	.7	.7		.3	.4	.3																	.3			
Front Crankshaft	19			.6		.8			.3					.8	.4	.8											
Oil Passage	20	.5		.2		.2	.7		.2		.1				.5	.6	.2	.4		.4				.1			.2
Flywheel	21								.5																		
Valve Retainer Nut	22	.8										.2													.4		
Cylinder Liner	23											.5							.6		.7						
Valve Guide	24						.1			.3																	
Cam	25	.3				.6																			.3		
Oil Pan	26					.8												.7			.5						

Fig. 7.9 DSM showing change impact of engine parts

initiating components in terms of change implementation to get the cost-effective CPP. In this case, the designer has two options, first to implement the change requirement in the component "Piston" design and then move to the component "Cylinder Block" design and vice versa. As the number of change requirements increases, the opportunities for the designer also increase to the find cost-effective propagation paths.

Table 7.2 Design duration of motorcycle engine components in days

S. no.	Design components	Design duration (days)
1.	Air valve	2
2.	Cylinder head	3
3.	Crankshaft pin	1
4.	Cylinder head gasket	1.5
5.	Crankcase	10
6.	Piston	5
7.	Gaseous ring	2
8.	Rear crankshaft	8
9.	Oil ring	4
10.	Piston pin	3
11.	Valve adjustable screw	2
12.	Strainer	2
13.	Connecting rod	7
14.	Camshaft	6
15.	Connecting rod journal	4
16.	Connecting rod bearing	4
17.	Engine oil pump	9
18.	Cylinder block	10
19.	Front crankshaft	8
20.	Oil passage	6
21.	Flywheel	5
22.	Valve retainer nut	2
23.	Cylinder liner	5
24.	Valve guide	7
25.	Cam	8
26.	Oil pan	3

7.6.1 Change Requirement Scenario

MCRs raised by numerous sources (internal/external) have different planned risks on the initiating components. Firstly, the designer investigated the initiating components on the basis of change requirements. After selecting the initiating components, the question arises that how to implement change requirement in an efficient manner. The aim of the designer is to find out the best possible sequence of the change requirements for implementation by exploring cost-effective CPP. From literature, it is revealed that in Britain new improvement has been observed that the Ministry of Defence as a customer is desirous in quantitative best-case, average-case, and worst-case risk evaluations. The same criteria have been implemented in this case study to explore the cost-effective CPPs.

Table 7.3 Categorization of planned risk

S. no.	Risk category	Risk value
1.	Low planned risk	$0 < \text{LPR} \leq 0.35$
2.	Average planned risk	$0.35 < \text{APR} \leq 0.7$
3.	High planned risk	$0.7 < \text{HPR} \leq 1$

For instance, two change requirements in motorcycle engine design are raised. One is to make changes in the design of connecting rod to sustain the pressure, which flows form the piston to the connecting rod. The other is to make changes in the design of cylinder head to maximize the flow of heat produced as a result of fuel combustion. The change in connecting rod and cylinder head design will propagate to the other parts of the engine such as piston, cylinder block, cylinder liner, crankcase and so on, because of the complex interdependencies between the engine components. The planned risks for both the initiating components have a range between 0 and 1, which depends on the severity of the change requirements. In this chapter, the planned risk associated with the change requirement is divided into 3 different categories, as highlighted in Table 7.3.

To find the optimal sequence of change requirements for execution, two different scenarios are considered. In the first scenario, it is assumed that both the change requirements have the same planned risk and in the second scenario, it is assumed that both have the different planned risk.

7.6.1.1 First Scenario

In this scenario, it is assumed that both the change requirements have the same planned risk. The planned risk is divided into three different categories, therefore in this scenario, three different cases will be analyzed and results will be compared with each other to find out the right sequence of change requirements for execution. The three different cases are named as:

- Both change requirements have low planned risk (Best-case scenario)
- Both change requirements have average planned risk (Average-case scenario)
- Both change requirements have high planned risk (Worst-case scenario)

The range of risk categories are defined in Table 7.3, and the low planned risk is taken as 0.2, for average planned risk is 0.5 and for high planned risk is 0.9. To implement the change requirement, the designer has two different options. In option 1, the designer will first implement change requirement in component 13 (Connecting rod) and move along the CPP obtained through the MCR-algorithm, and when the component 2 (Cylinder head) arrive in the propagation path, the change requirement is implemented in that component (Cylinder head). The second option is contrary to the earlier.

When the MCR-algorithm is executed for all the three scenarios via option 1 and option 2, different potential solutions are available to implement the change requirements in the engine design as shown in Fig. 7.10. In all the three scenarios

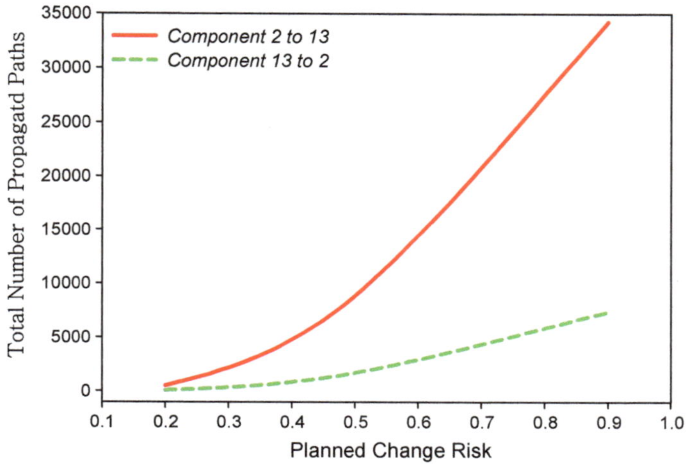

Fig. 7.10 Total number of propagated paths for three different case scenarios

the number of CPPs in option 2 is almost 5 times than option 1. It is obvious from Fig. 7.10 that the number of propagated paths increases exponentially as the planned risk increases for the change requirement. When the planned change risk increases, more and more potential solution are available to resolve the change requirements but also the TCT required to implement the changes may also increase. The minimum completion time and maximum completion time of different propagated paths are lower in option 1 as compare to option 2, for all the case scenarios as shown in Figs. 7.11 and 7.12. In the case study, design component completion time and the differences between design completion times of various components are quite small, as a result, CPPs with a short duration differences are achieved. Numerous CPPs can be attained if the design component duration and the differences in design component duration are larger in the design process model.

If all the CPPs for option 1 and option 2 are considered in three distinct case scenarios, it can be concluded that the number of change step increases with increase in planned change risk and also by changing the sequence of the change requirement. In the best-case scenario, there are 57.45% of CPPs, which can go up to 9 numbers of changes (steps) in option 1 and 21.56% of CPPs, which can go up to 10 numbers of changes (steps) in option 2, as depicted in Fig. 7.13. In the average-case scenario, the number of changes (step) reaches up to 13 for option 1 and 14 for option 2, as shown in Fig. 7.14. Similarly, it can be seen in Fig. 7.15 that in a worst-case scenario the number of changes (step) reaches up to 14 in option 1 and 15 in option 2, which are 3.78% and 0.17% respectively of all the propagated paths in both options.

The propagation path with more distinct components will have longer completion time to resolve the change requirements. The number of distinct components also depends upon the sequence of change requirements, in which they are executed

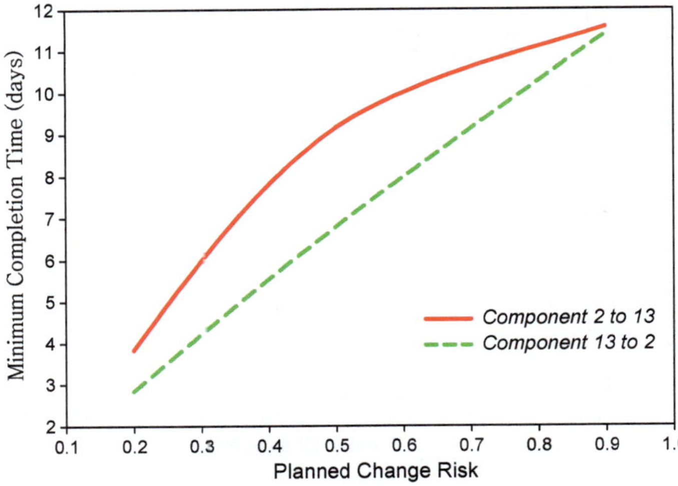

Fig. 7.11 Minimum completion time for three different case scenarios

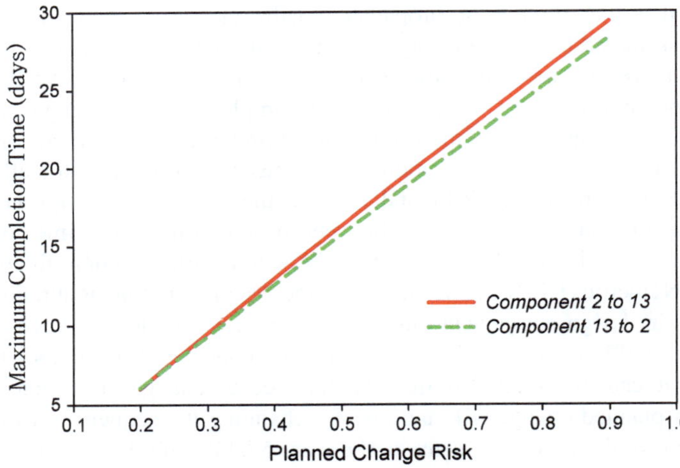

Fig. 7.12 Maximum completion time for three different case scenarios

and the planned change risk. In all case scenarios, it is evident from Figs. 7.16, 7.17 and 7.18 that in option 1 the number of distinct components is less than in option 2. It is evident from all figures that some design components are iterated 2–3 times in the CPP. The path which contains a smaller number of components and change steps, it will be easier and have a less time to implement the changes.

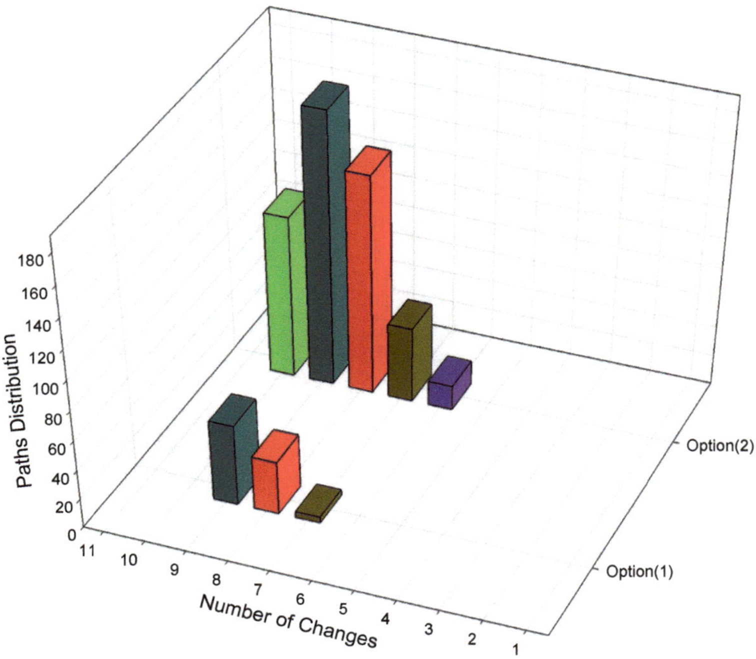

Fig. 7.13 Number of changes in best-case scenario

Several CPPs are available as a result of MCR-algorithm execution for a number of change requirements. In the proposed case study, the minimum numbers of CPPs are 94 in a best-case scenario for option 1, and maximum numbers of CPPs are 34,293 in a worst-case scenario for option 2. To analyze the design duration of the entire set of propagated paths for all the scenarios, a box plot is presented in Figs. 7.19, 7.20 and 7.21.

The box plot provides the deeper insight of the available data and is used to visualize the spread, distribution, and location of the represented data. It is evident from Fig. 7.19 the in best-case scenario; the data is positively skewed in option 1, which means that data points are closer to each other in lower half while the data in upper half has a widespread. Whereas, in option 2 the data is slightly negatively skewed, which means that data points are concentrated in the upper portion. The median of data set in all scenarios for option 1 is lower than option 2, which means that more than 50% of CPPs in option 1 have less design duration than 50% in option 2. Other aspects of CPPs design duration for all the three case scenarios are presented in Tables 7.4, 7.5 and 7.6.

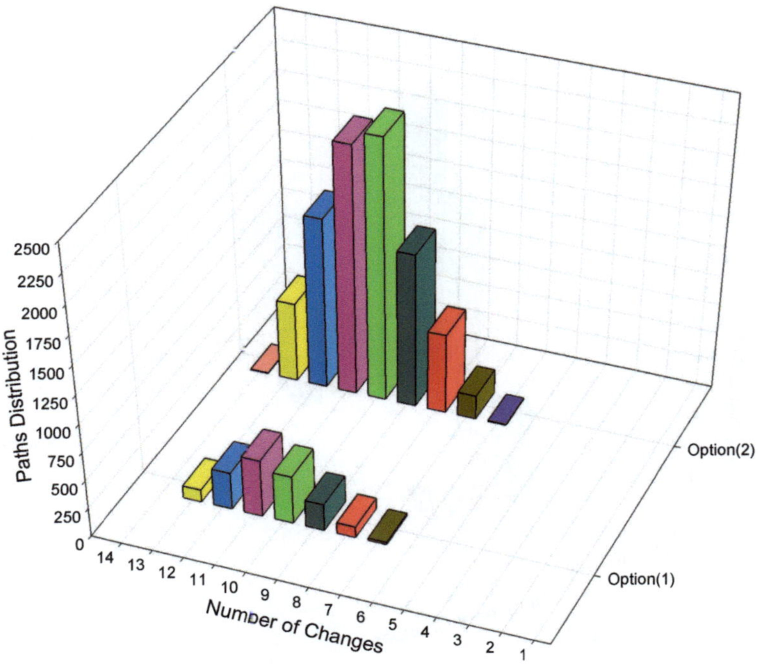

Fig. 7.14 Number of changes in average-case scenario

7.6.1.2 Second Scenario

In this scenario, it is assumed that both the change requirements have different planned risk. According to planned risk, six different scenarios are analyzed to implement two change requirements. The six different scenarios are named as:

- One change requirement has high planned risk, and other has low planned risk and vice versa
- One change requirement has high planned risk, and other has average planned risk and vice versa
- One change requirement has average planned risk, and other has low planned risk and vice versa

After execution of MCR-algorithm, many solutions are available in all the cases to implement the change requirements. Minimum completion time can be defined as "the least time required applying the change requirements in any set of possible solutions for specific scenario". It is evident from Fig. 7.22 that the minimum completion time for option 1 in all scenarios is less than option 2.

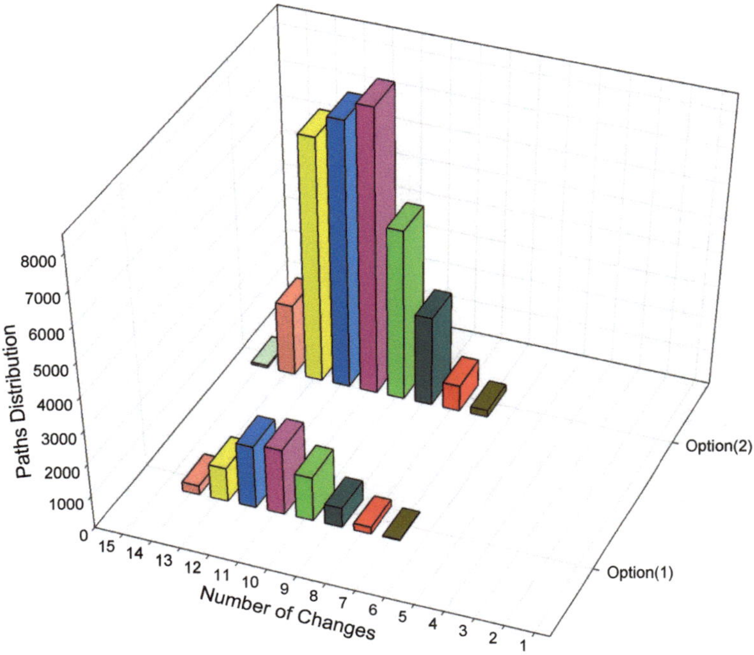

Fig. 7.15 Number of changes in worst-case scenario

For instance, if the change requirement 1 has a planned risk of 0.9 on component 13 and the change requirement 2 has a planned risk of 0.5 on component 2. The designer has two possibilities for the execution of change requirements, either start implementation of change requirement from component 13 and move to component 2 "$C_{13}(0.9)$–$C_2(0.5)$" or vice versa "$C_2(0.5)$–$C_{13}(0.9)$". The red line in Fig. 7.22 highlights the example discussed and it can be seen that minimum completion time in option 1 (C_{13}–C_2) is 9.83 days and in option 2 (C_2–C_{13}) is 12.28 days. It can be seen that in some cases the slope of lines is very steep as compare to others, it means that there is a large difference in the minimum completion time of both options.

Similarly, maximum completion can be defined as "the greatest time required implementing the change requirements in any set of possible solutions for specific scenario". Therefore, the difference between maximum completion time and minimum completion time defines the range of the completion time for all the possible solutions in a specific scenario. With reference to Fig. 7.23, it can be seen that in all cases except 2, the maximum completion time of option 1 is less than option 2.

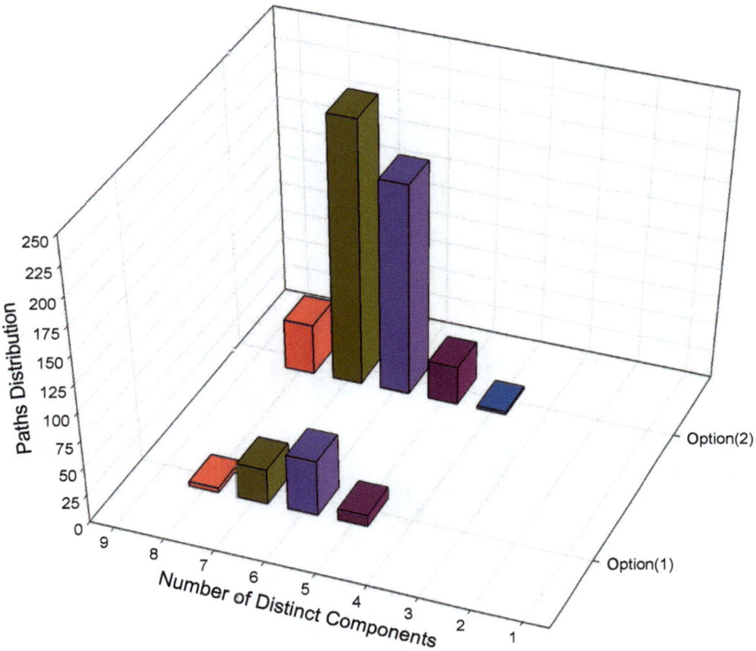

Fig. 7.16 Number of distinct components in best-case scenario

For instance, the maximum completion time for option 1 [C_{13} (0.2)–C_2 (0.9)] is 15.148 days and for option 2 is 16.31 days, as highlighted by pink line in Fig. 7.23. In the proposed case study, the differences between design completion times of various components are quite small, as a result, CPPs with a short duration differences are obtained. Numerous CPPs are available in option 1 and option 2 for the implementation of MCRs as shown in Tables 7.7 and 7.8. The maximum numbers of CPPs in option 2 are 19,453 when the planned risk for component 2 is 0.5 and 0.9 for component 13. When the change requirements have planned risk of Low–High and Low–Average, there is a huge difference in the number of CPPs for both options as evident from Tables 7.7 and 7.8.

7.7 Conclusion and Future Work

In this chapter, MCR-algorithm is suggested, which can be used to predict the different CPPs as a result of distinct change requirement received from various sources. In a complex product design, all the components are connected to each

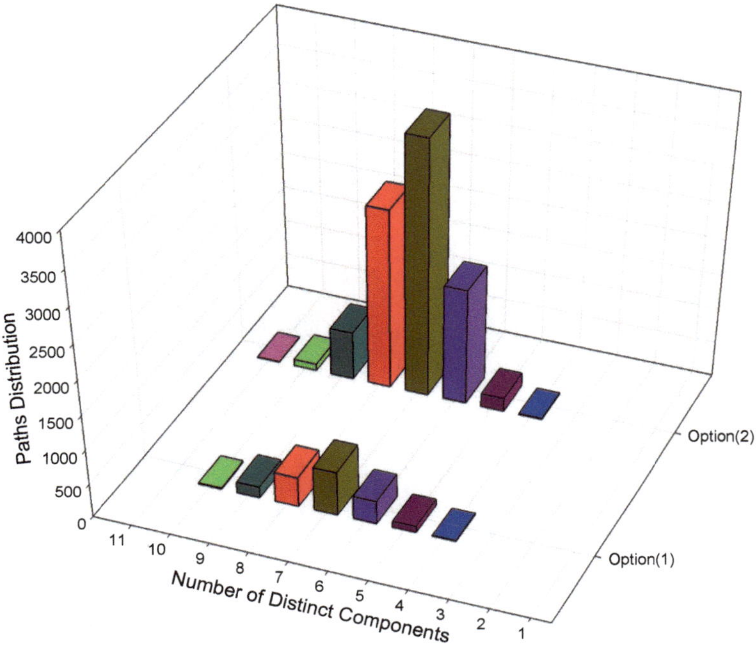

Fig. 7.17 Number of distinct components in average-case scenario

other in a closed fashion; therefore, ECs have high possibility to propagate throughout the entire product's structure. EC requirements have different causes to be initiated. The repercussions of such change requirements are very severe if they are not managed accurately.

The main emphasis in ECM is to predict the effects of change requirements and manage these effects so that the design process can be wrapped up on time with minimum cost. Change requirements are the prime risk to the PD processes, which mostly leads to extended schedule and cost overrun. Increase in product complexity with a lesser development time makes it essential to implement the MCRs in an appropriate way. Obviously, it is an ideal situation to achieve the right design for the first time, removing the need to do changes. Though, ECs due to evolving customer requirements and/or wrong decisions are unavoidable. It is worth mentioning that ECM plays a vital role in successful PD process.

The author introduces a novel approach to handle MCRs in an efficient way. In this chapter, a process modeling and DSM-based technique is applied. It investigates the cost-effective CPPs and provides the best possible sequence of MCRs, in which they should be executed in the design process. The proposed MCR-algorithm

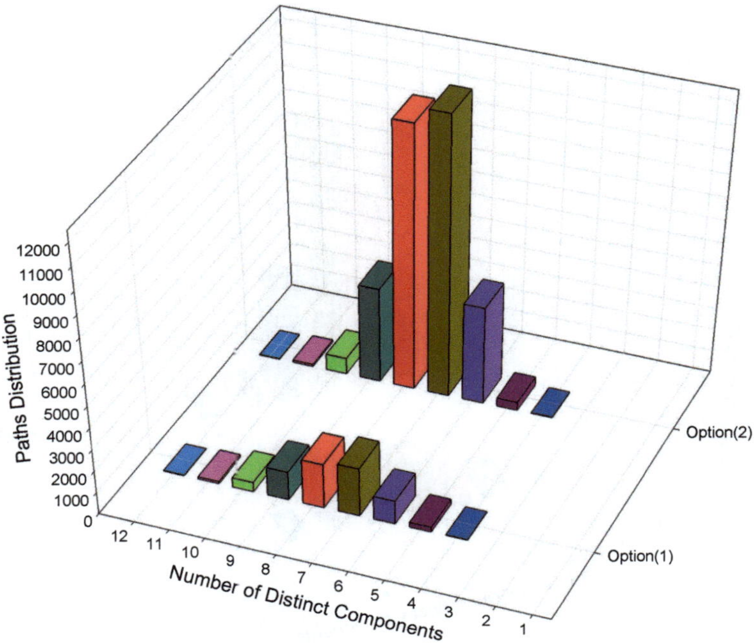

Fig. 7.18 Number of distinct components in worst-case scenario

Fig. 7.19 Design duration distribution in best cases scenario

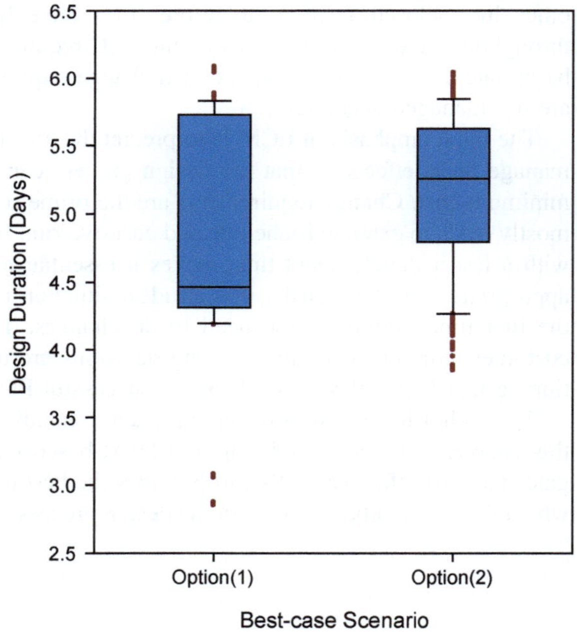

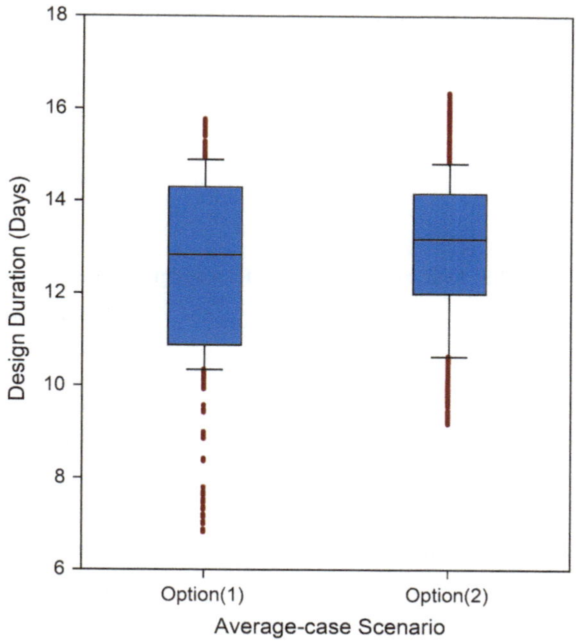

Fig. 7.20 Design duration distribution in average cases scenario

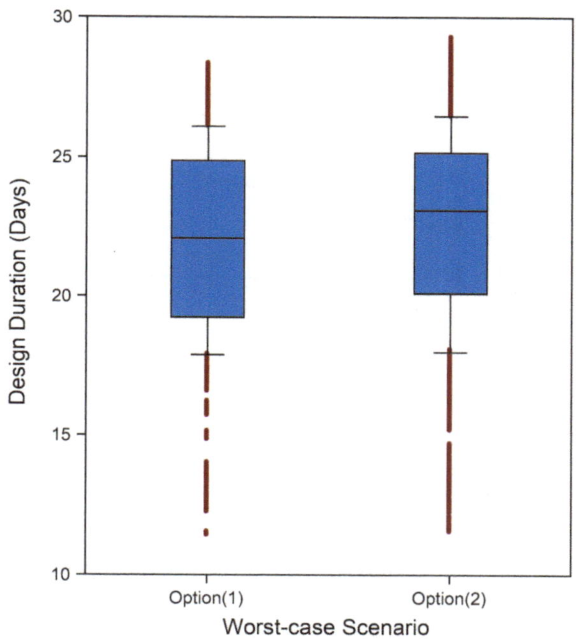

Fig. 7.21 Design duration distribution in worst cases scenario

Table 7.4 Different aspects of CPPs redesign duration (day) in best-case scenario

S. no.	Description	Best-case scenario	
		Option 1	Option 2
1.	Range	3.232	2.191
2.	Median	4.467	5.261
3.	Mean	4.847	5.166

Table 7.5 Different aspects of CPPs redesign duration (day) in average-case scenario

S. no.	Description	Average-case scenario	
		Option 1	Option 2
1.	Range	8.951	7.157
2.	Median	12.837	13.169
3.	Mean	12.557	12.960

Table 7.6 Different aspects of CPPs redesign duration (day) in worst-case scenario

S. no.	Description	Worst-case scenario	
		Option 1	Option 2
1.	Range	16.933	17.815
2.	Median	22.143	23.32
3.	Mean	21.907	22.764

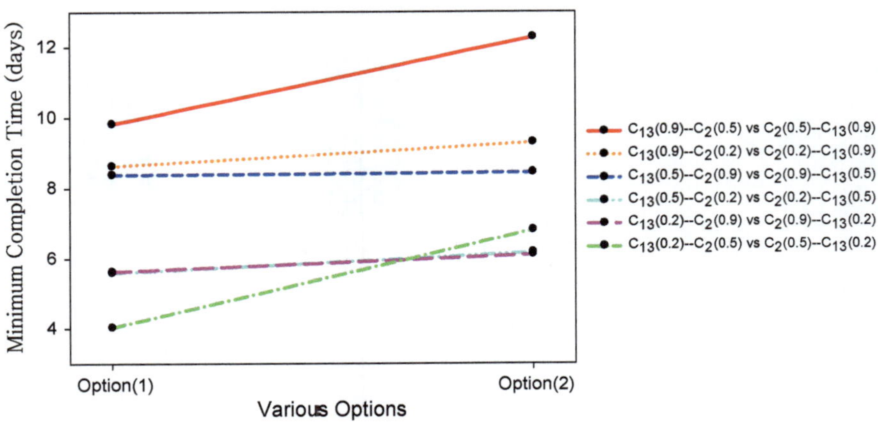

Fig. 7.22 Minimum completion time of six different scenarios

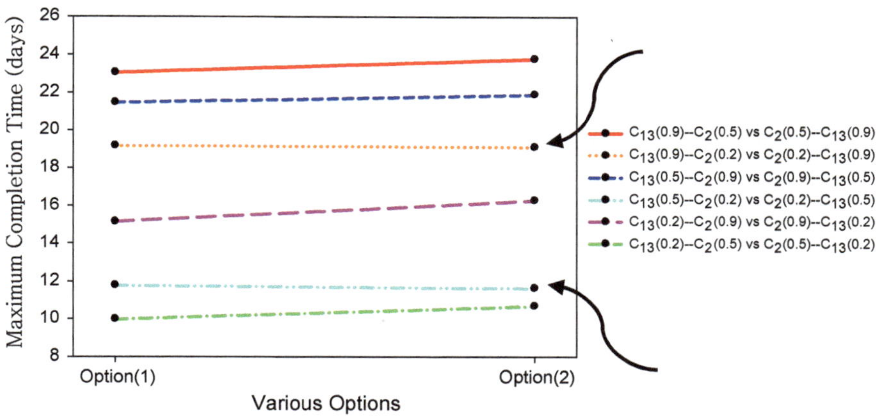

Fig. 7.23 Maximum completion time of six different scenarios

Table 7.7 Number of propagation paths in option 1

S. no.	Option 1	No. of paths
1.	$C_{13}(0.9)$–$C_2(0.5)$	4253
2.	$C_{13}(0.9)$–$C_2(0.2)$	1393
3.	$C_{13}(0.5)$–$C_2(0.9)$	2928
4.	$C_{13}(0.5)$–$C_2(0.2)$	556
5.	$C_{13}(0.2)$–$C_2(0.9)$	499
6.	$C_{13}(0.2)$–$C_2(0.5)$	297

Table 7.8 Number of propagation paths in option 2

S. no.	Option 2	No. of paths
1.	$C_2(0.5)$–$C_{13}(0.9)$	19,453
2.	$C_2(0.2)$–$C_{13}(0.9)$	4086
3.	$C_2(0.9)$–$C_{13}(0.5)$	15,603
4.	$C_2(0.2)$–$C_{13}(0.5)$	1846
5.	$C_2(0.9)$–$C_{13}(0.2)$	4392
6.	$C_2(0.5)$–$C_{13}(0.2)$	2434

provides risk propagated to the entire system as a result of planned risk, thus providing designers with an excellent overview of the product. The planned risk is divided into three different categories named as, low, average and high risk. In this research work, planned risk and propagated risk only refers to the product's components redesign cost.

A case example is used to demonstrate the practicality of the proposed method. Motorcycle engine design and two change requirements with different planned risks, for the validation of suggested MCR-algorithm, are considered. All possible scenarios of change requirements with respect to planned risk are assessed. It is evident from the results that different arrangements have different outcomes, such as; the number of CPPs, the number of change steps, the number of distinct components and design duration. The study showed that in MCRs the execution sequence of different change requirements has a great impact on the results. The results propose that the method developed is appropriate for assessing different cost-effective CPPs.

References

1. Otto K, Wood K (2001) Product design: techniques in reverse engineering, systematic design and new product development. Prentice-Hall, New York
2. Jarratt TAW, Eckert CM, Clarkson, PJ (2006) Pitfalls of engineering change: change practice during complex product design. In: Advances in design, 1st edn. Springer Series in Advanced Manufacturing, Germany, 413–424
3. Evans S (1990) Implementation framework for integrated design teams. J Eng Des 1(4):355–363
4. McMahon C, Busby J (2005) Risk in the design process. In: Clarkson J, Eckert C (eds) Design process improvement: a review of current practice. Springer, London, pp 286–305
5. Clarkson PJ, Simons C, Eckert C (2004) Predicting change propagation in complex design. J Mech Des 126(5):788–797
6. Eckert CM, Clarkson PJ, Zanker W (2004) Change and customisation in complex engineering domains. Res Eng Des 15(1):1–21
7. Suh ES, de Weck OL, Chang D (2007) Flexible product platforms: framework and case study. Res Eng Des 18(2):67–89
8. Lee H et al (2010) An analytic network process approach to measuring design change impacts in modular products. J Eng Des 21(1):75–91
9. Yang F, Duan GJ (2012) Developing a parameter linkage-based method for searching change propagation paths. Res Eng Des 23(4):353–372
10. Hamraz B, Caldwell NHM, Clarkson PJ (2013) A matrix-calculation-based algorithm for numerical change propagation analysis. IEEE Trans Eng Manag 60(1):186–198
11. James E, Kelley J, Walker MR (1959) Critical-path planning and scheduling. In: Eastern joint IRE-AIEE-ACM computer conference. ACM, Boston, pp 160–173
12. Malcolm DG et al (1959) Application of a technique for research and development program evaluation. Oper Res 7(5):646–669
13. Slyke RMV (1963) Monte Carlo methods and the PERT problem. Oper Res 11(5):839–860
14. Elmaghraby SE, Kamburowski J (1992) The analysis of activity networks under generalized precedence relations (GPRs). Manage Sci 38(9):1245–1263
15. Neumann K, Steinhardt U (1979) GERT networks and the time-oriented evaluation of projects. Springer, New York, p 268
16. Steward DV (1981) The design structure system: a method for managing the design of complex systems. IEEE Trans Eng Manag EM-28(3):71–74
17. Eppinger SD et al (1994) A model-based method for organizing tasks in product development. Res Eng Des 6(1):1–13
18. Ulrich KT, Eppinger SD (1995) Product design and development, 1st edn. McGraw-Hill, New York
19. Sanderson AC, Zhang H, Mello LSH (1990) Assembly sequence planning. Ai Mag 11(1):62–81

20. Belhe U, Kusiak A (1996) Modeling relationships among design activities. J Mech Des 118 (4):454–460
21. Deyi X (1997) A multilevel optimization approach considering product realization process alternatives and parameters for improving manufacturability. J Manuf Syst 16(5):337–351
22. Simon H (1996) The sciences of the artificial. MIT Press, Cambridge
23. Cohen T, Navathe SB, Fulton RE (2000) C-FAR, change favorable representation. Comput Aided Des 32(5–6):321–338
24. Jarratt T, Eckert C, Clarkson PJ (2004) Development of a product model to support engineering change management. In: Horvath I, Xirouchakis P (eds) Tools and methods of competitive engineering, vol 1 and 2. Mill press, Lausanne, Switzerland, pp 331–342
25. Martin MV, Ishii K (2002) Design for variety: developing standardized and modularized product platform architectures. Res Eng Des 13(4):23
26. Ollinger GA, Stahovich TF (2004) RedesignIT—a model-based tool for managing design changes. J Mech Des 126(2):208–216
27. Koehler C et al (2008) A matrix representation of the CPM/PDD approach as a means for change impact analysis. In: Marjanovic D et al (eds) Proceedings of the 10th international design conference, pp 167–174
28. Eckert CM et al (2006) Supporting change processes in design: complexity, prediction and reliability. Reliab Eng Syst Saf 91(12):1521–1534
29. Smaling R, de Weck O (2007) Assessing risks and opportunities of technology infusion in system design. Syst Eng 10(1):1–25
30. Tang D et al (2008) Design structure matrix-based engineering change management for product development. Int J Int Manuf Serv 1(3):231–245
31. Giffin M et al (2009) Change propagation analysis in complex technical systems. J Mech Des 131(8):081001-081001-14
32. Shankar P, Morkos B, Summers JD (2012) Reasons for change propagation: a case study in an automotive OEM. Res Eng Des 23(4):291–303
33. Morkos B, Shankar P, Summers JD (2012) Predicting requirement change propagation, using higher order design structure matrices: an industry case study. J Eng Des 23(12):902–923
34. Pahl G, Beitz W (1998) Engineering design: a systematic approach. Springer, London
35. Ullman D (2003) The mechanical design process. McGraw-Hill, New York
36. Hull E, Jackson K, Dick J (2005) Requirements engineering, 2nd edn. Springer, London
37. Kotonya G, Sommerville I (1992) Viewpoints for requirements definition. Softw Eng J 7(6): 375–387
38. Hongxing C et al (2007) Requirements change management of information system based on the keyword mapping. In: 6th Wuhan international conference on E-business, management challenges in a global World, vol 1–4, pp 135–140
39. Van Bossuyt DL et al (2013) On measuring engineering risk attitudes 1. J Mech Des 135 (12):121001/1–121001/13
40. Zhang J et al (2015) Optimisation of critical chain sequencing based on activities' information flow interactions. Int J Prod Res 53(20):6231–6241
41. Hollins B, Pugh S (1990) Successful product design. Butter-worths, London
42. Maier JF et al (2014) Simulating progressive iteration, rework and change propagation to prioritise design tasks. Res Eng Des 25(4):283–307
43. Andersson J, Pohl J, Eppinger SD (1998) A design process modeling approach incorporating nonlinear elements. In: ASME 10th international conference on design theory and methodology, Atlanta, GA
44. Smith RP, Eppinger SD (1997) A predictive model of sequential iteration in engineering design. Manage Sci 43(8):1104–1120
45. Smith RP, Morrow JA (1999) Product development process modeling. Des Stud 20(3):237–261
46. Tang D et al (2010) Analysis of engineering change impacts based on design structure matrix. J Mech Eng 46(01):154–161
47. Li Y, Zhao W, Shao X (2012) A process simulation based method for scheduling product design change propagation. Adv Eng Inf 26(3):529–538

Chapter 8
Product-Oriented Change Propagation and Prediction Approach in Product Family Design

8.1 Introduction and Synopsis

The change propagation and prediction in a product family design is discussed in this chapter. In the present situation, product family design is a popular approach in the industry to meet the diversified design needs and customer preferences. In a product family, common components are used by different product variants. The changes in one product variant will flow to the entire family. Therefore, the prediction of change propagation in a family of the product is a challenging task. In the literature, various design change management methodologies have been proposed, but all these methods are limited to a single artifact.

In this chapter, first, the definitions and terminologies used in the product family are elaborated with the pictorial representation. After that, the seven steps state-of-the-art approach to explore the cost-effective CPPs in a product family is presented. The propagated risk is quantified in terms of change impact and propagation probability. Component-based DSM, mathematical model and an algorithm are proposed to investigate the change propagation across the family of products. Finally, in this chapter, the product family of an electric kettle is presented as an industrial case study to demonstrate the implementation of the advanced approach. Different CPPs affecting distinct product variants are achieved with different redesign time. The outcomes reveal that the proposed technique is appropriate for analyzing different CPPs in a product family with an aim to minimize the cost of redesign.

Nowadays, industrialists are facing a challenging environment due to clients' ever-changing expectations and complex product design structure. To cope with such demanding situations, most of the firms have changed their manufacturing strategy. In the past organizations were mainly focused on the individual product but with a passage of time, the attention diverted towards product families [1]. In today's environment, product family design is a dominant strategy in the industry to meet the diversified design requirements and customer preferences. Product family

D. Tang et al., *Matrix-Based Product Design and Change Management*,
DOI 10.1007/978-981-10-5077-0_8

offers a broad range of artifacts. However, the basic functionality remains the same across the products within a family. A product family can be defined as "a set of distinct products, which share some common elements, modules or sub-systems and functions" [2, 3]. Each product within the family is termed as a product variant, and it has distinctive specifications to meet diverse requirements. The common elements are usually defined as the product platform [4]. Factors such as shorter product lifecycle, changing market needs, capture new market niches, making it essential for an organization to improve existing product families as much as possible. To provide a variety of products at minimal cost, product family design and platform-based PD are efficient approaches [5, 6].

Firms can target wider market opportunities owing to product family approach. On the other hand, product family also adds complications in the design process due to shared elements. The product family is used to meet the requirements of potential customers and to enter new market niches. New requirements or modifications can influence the operational features and structure of the product to be modified as well as on the other family members that share common elements. Therefore, the prediction of change propagation in a family of the product is a challenging task. Numerous design change management tools and methodologies have been proposed in the literature to evaluate the impact of ECs on the product [7–9], but they are restricted to a single product instance. This shortcoming focusses our attention on the expansion of ECM techniques. A method should be developed to evaluate the impact of alterations, which propagates from a single artifact to the whole product family.

In view of the above, this chapter presents a framework, which explores effective CPPs by considering the risk associated with design changes in the product family, with an aim to minimize the overall cost of redesign. Quantitative risk analysis based approach is utilized to evaluate the propagated risk from the initiating element in a single product to the entire family, as a result of change requirement. Risk is defined as "the likelihood of occurrence of an incident times the strength of the outcomes" [10]. Risk propagates due to the information flow between directly connected elements and leads to rework in projects [11]. Rework usually causes project delays and cost overrun. The propagated risk, which would result in rework, is quantified in terms of change impact and change propagation probability. DSM based mathematical model and an algorithm are proposed to investigate the change propagation across the family of products. The development of a product redesign process is described by 'AND/OR' logical relationships between product elements. Finally, to demonstrate the implementation of advanced mathematical model and algorithm, a case study of an electric kettle product family is considered. The outcomes reveal that the proposed technique is appropriate for evaluating different CPPs in a product family and also provides some additional information, i.e., the number of design elements and change steps involved in change propagation. The proposed technique would give a better insight to estimate the total cost of redesign in terms of lead time in a product family.

8.2 Engineering Change in Product Family Design

In recent decades, the customer's requirements continuously evolve, and their need for a variety of products has escalated rapidly. Many firms are trying to offer more product variants to fulfill the growing demand in the existing and new market niches without interrupting the production efficiency [12, 13]. Product family design and platform-based PD have become most popular manufacturing techniques, to cope with the ever-changing customer's expectations. To satisfy the client's need by providing a variety of artifacts increases firm's sale. However, producing a large number of product variants add complications in the process, thus increases cost and the decline is observed in a company's profit [14]. Therefore, the trade-off should be maintained between the product variety and process complexity ensuing from product diversity to realizing the optimal results [15]. A product family can be construed as a group of similar products with unique features that share some common elements as illustrated in Fig. 8.1.

The aim of the product family is to meet specific client needs. The elements in a product family can be categorized into three different groups, namely, the family platform, product platform and sole components. The elements that are part of all the product variants in the family are known as a family platform, i.e., element 8. The components, which exist in more than one product variant, but not in the entire family, are known as product platform, i.e., elements 5, 6 and 7. The element that exists only in a single product variant is termed as a sole component, i.e., elements 1 and 2. According to Meyer and Lehnerd, family and product platforms are the

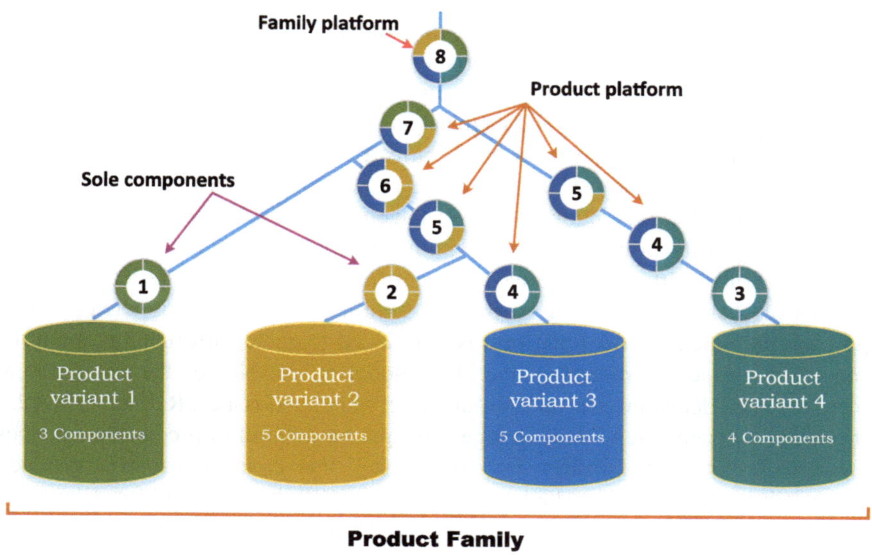

Fig. 8.1 Product family comprising of four product variants

collections of sub-systems and interfaces, which construct a common structure from which a series of varied products can be effectively manufactured [3]. The size of a product family increases owing to new requirements, to capture new market segments and potential customers. The assessment of the new product features regarding resources, time, cost, and environmental impact must be evident to design manager. New requirements and also altering the existing ones affect many aspects of the product. Therefore, the designer must be equipped with tools and methods to evaluate the impact of these changes easily. For that purpose, different tools and methods have been suggested in the literature to highlight the consequences of ECs on a product [16, 17].

ECs have an important role in the PD process. Mostly, new products are developed through modifications of the existing ones. Change propagation management based on minimal cost is very significant in the product designing. The unrestrainable propagation of a simple EC in a system instigates a burst of changes in the entire system and eventually results in system collapse [18]. Change Prediction Method (CPM) is a technique to predict change propagation risks, which are based on product linkage models [17]. The authors claimed that CPM method provides a good indication of future change likelihood without the need for detailed knowledge of the PD process. Keller et al. explored possible CPPs using CPM, which permits avoidance of propagation in the future by blocking off risky CPPs [19]. The focus of this research was on the incremental products where changes are made only in those components of the product, which take part in the change propagation while other components remained unchanged. Ahmad et al. detected the potential change propagation linkages to evaluate the change impact [7]. To discover CPPs, Yang and Duan developed a Parameter Linkage Network approach [20]. Based on functional and physical models, a matrix-based technique was established by Fei et al. [21] to investigate change propagation between product's components, which assist designers to explore design conflicts arising from design changes.

From the literature, it is evident that previous studies have mainly focused on change propagation prediction and its management in a single artifact. Considerable research on product family design has been carried out over the last two decades. The literature covers almost all aspects of product family designing and manufacturing except EC prediction and its management. Little consideration has been given to EC propagation management in a product family. An approach, known as active distributed virtual change environment (ADVICE) was suggested to deliver the idea of change propagation in product families [22]. It is a knowledge-based approach, in which a virtual platform is introduced to gather information about the process to reduce the redesign time. The model is suitable for the environment where repeated decisions are made around a set of defined issues. Raffaeli et al. [23] introduced a dependency-based change propagation method to predict the changes extended to product families by considering the concept of meta-product. Meta-product is fictitious product and encompasses all the parts and functionalities of all the product variants in the family. Based on the above discussion, we propose a quantitative risk-based analysis approach to assess the change propagation across the product family.

8.3 State-of-the-Art Approach

8.3.1 Change Propagation Mechanism in Product Family

The proposed approach allows the designers to trace the CPPs initiated as a result of design requirements to affect the elements of product variants in the family. The main focus of the advanced technique is to satisfy the change requirement in a cost-effective way and to accomplish the project without delays. In the suggested methodology, when the propagated risk is less than 1% of the initial design effort, the propagation process should be stopped, as the risk is minuscule and may be neglected. It is termed as threshold risk. The complexity of the problem increases when the changes propagate across numerous product variants in the family. The propounded method proceeds in seven stages as illustrated in Fig. 8.2. To specify this technique, different steps are detailed as follows:

Step 1: Engineering design changes in a product are essential due to global competition in the market and also to rectify the errors incorporated during the initial design due to lack of information or misinterpretation of a customer's requirements. Changes can occur in any phase of the product lifecycle. The system components can be affected by the changes, whether it is desirable or undesirable. ECs are the primary source of originating risk, in terms of time, cost and quality. A single EC in any product in the family can influence all other products in the family because of shared elements. Numerous sources can instigate changes in product design, i.e., performance improvement, error rectification, quality issues, manufacturing constraints, changes in the customer's specifications, to capture new

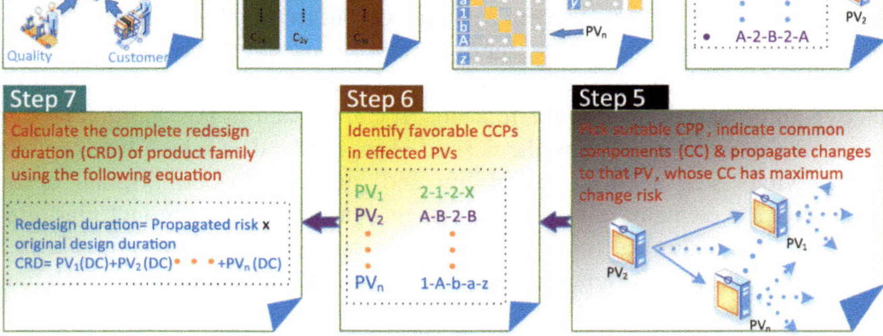

Fig. 8.2 Proposed approach for change prediction in a product family

market niches and potential clients, current legislation, and safety concerns. As a result, change requirement is initiated in the design process to fulfill the need.

Step 2: Product is primarily made up of several sub-systems, sub-assemblies, components and parts, which are arranged in such a fashion that product's function is assigned to the physical components. Product family consists of several product variants and each product variant comprises of different components. After the initiation of change requirement, all product variants in the family are analyzed in the second step. For that purpose, all product variants must be broken down into sub-assemblies or components, which is the first phase of constructing product model. Thus, each product variant can be viewed as a collection of elements. Designers and engineers acquainted with the original design of the product can suggest the extent, to which the product must be broken down.

Step 3: The product decomposition into elements permits interconnectivity within the artifact design to be viewed in a DSM. A component-based DSM approach is applied in the proposed method. DSM is a modeling tool to represent the system elements and their interactions in a simple and compact manner. Steward introduced DSM to model the product, which provides simplicity and conciseness in the representation of a complex data [24]. Browning in his study declared that 1000 DSM articles had been cited till November 2015 [25]. The second phase of product model construction is to investigate the direct dependencies between the elements regarding likelihood and change impact. This activity has been accomplished by interviewing experience designers of the related field. The likelihood and change impact vary between the components owing to the interfaces through which they are connected such as geometrical joints, material flow, electrical connections and the flow of information that take place between the components. In product family each product variant has a separate DSM as shown in Fig. 8.2. The elements specify the rows and columns heading in the DSM-based product model. The diagonal cells are usually empty, which illustrate that reflexive relation is overlooked, or sometimes exhibit the component duration. The off-diagonal cells denote the relation between the system elements, in which numeric values are used to define the change impact and likelihood. An off-diagonal cell of DSM in the ith row and jth column denotes the relationship that the ith row element affects the jth column element, which means that the jth column element receives information from the ith row element and the ith row element sends its output to the jth column element.

Step 4: As we know that product family comprises of several product variants. In this step, product variant in which changes should be incorporated is selected as per change requirement received in the first step. It can be seen in Fig. 8.3 that product variant 1 is chosen to implement the desired change. The data which was captured in the previous step such as probability and impact matrices, original design duration of each element, common components (a family and product platform), and risk threshold are the input to the simulation model as shown in Fig. 8.3. Designers will select the originating element based on their experience, which must

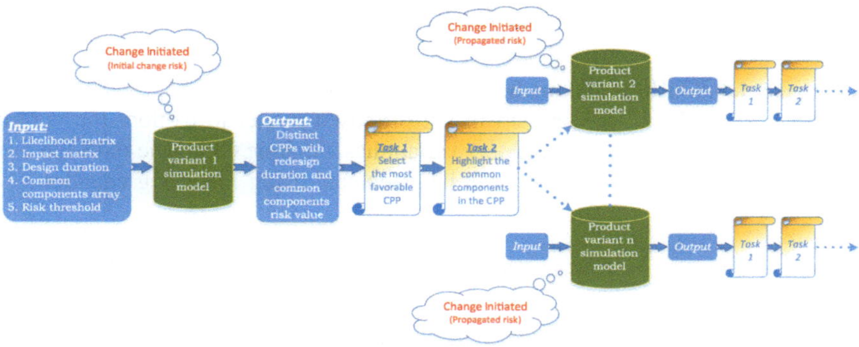

Fig. 8.3 Change flow pattern in product family

be modified to incorporate the change request. The value of probability in the case of the originating component is 1, and the value of change impact depends on the severity of the change requirement. Different CPPs are explored using the proposed mathematical model and advanced algorithm. Each propagation path has a redesign time and the list of common components involved in the change propagation along with the change risk.

Step 5: During the change propagation process different solution alternatives are usually available to meet the change requirements. Each design solution is distinguished from others by distinct design components involved in the change propagation process. Suitable CPP is selected by designers, and the common elements (a family and product platform) are highlighted as depicted in Fig. 8.3. The common components involved in the change propagation process have different change risks. Some common components may be iterated in the propagation path. Therefore, the total change risk of the iterated common components is computed by adding the individual risks. The common component with maximum change risk is selected. The design changes will propagate to that product variant whose common components have maximum change risk. This change risk will serve as the initiating risk for the product variant.

Step 6: In the same way changes will propagate to different product variants in the family. The change propagation will continue until the change risk reaches the defined threshold value. Each product variant has several CPPs. Designers select the appropriate CPP from each product variant to incorporate the design changes.

Step 7: In the last step, the solo components and the common components with a maximum change risk is selected from each CPP. The change risk of each element is multiplied by its original design time to calculate the redesign time of concerned component. The TCT of product family is calculated by adding all the common components redesign time from multiple product variants.

8.3.2 Change Propagation Pattern and Mathematical Model

The logical relationships and mathematical model described in Chap. 7, Sects. 7.4.3 and 7.5.1 respectively, are applied in the proposed model to explore the various CPPs. In product family, each product variant has a unique CPP. The redesign time to accomplish the design changes can be calculated using following equation.

$$RD_{PV_x} = \sum_{i=1}^{k_x} PR_{i-1,i} \times DT_{C_i} \qquad where \quad x = 1, 2, 3, \ldots\ldots, n \qquad (8.1)$$

In Eq. (8.1), 'x' represents the product variant in a family of the product and 'DT_{Ci}' is the design time of the component 'C_i' in hours. The TCT of product family can be calculated by adding the redesign time of each product variant, considering the maximum redesign time of common components only once.

8.4 Advanced Algorithm for Change Prediction in Product Family

In this section, an advanced algorithm to explore the effective CPPs in product family using the aforementioned mathematical model and logical relationships is introduced. The flowchart of the proposed algorithm is illustrated in Fig. 8.4.

Let's consider a product family comprising of 'n' products variants. The propagation probability and change impact matrices of 'n' products variants, design duration of each component, a common component array (CCA) and risk threshold are the input of the proposed algorithm. Three data arrays, named as, finished path array, in-process path array and common component risk array (CCRA) are established in the algorithm. Finished path array is used to save the completed paths with their redesign duration and common components with their propagated risk, while in-process path array can store the uncompleted propagation paths during the simulation process.

The in-process path array stores the propagation paths in the form of layers, and it serves as FIFO queue. In each simulation step, all the child components accompanied by their parent components are stored in successive layers. CCRA is used to store the common elements along with the propagated risk, which lies in the CPP during the simulation process. When the path is completed, it will be stored in the finished path array along with the common components and their propagated risk.

Two types of possibilities can exist in the simulation process. In the case of a single change requirement, when the component array is selected from the in-process path array, the components will be checked for iteration as well as the presence of an array final element in the CCA. After that, the propagated risk will

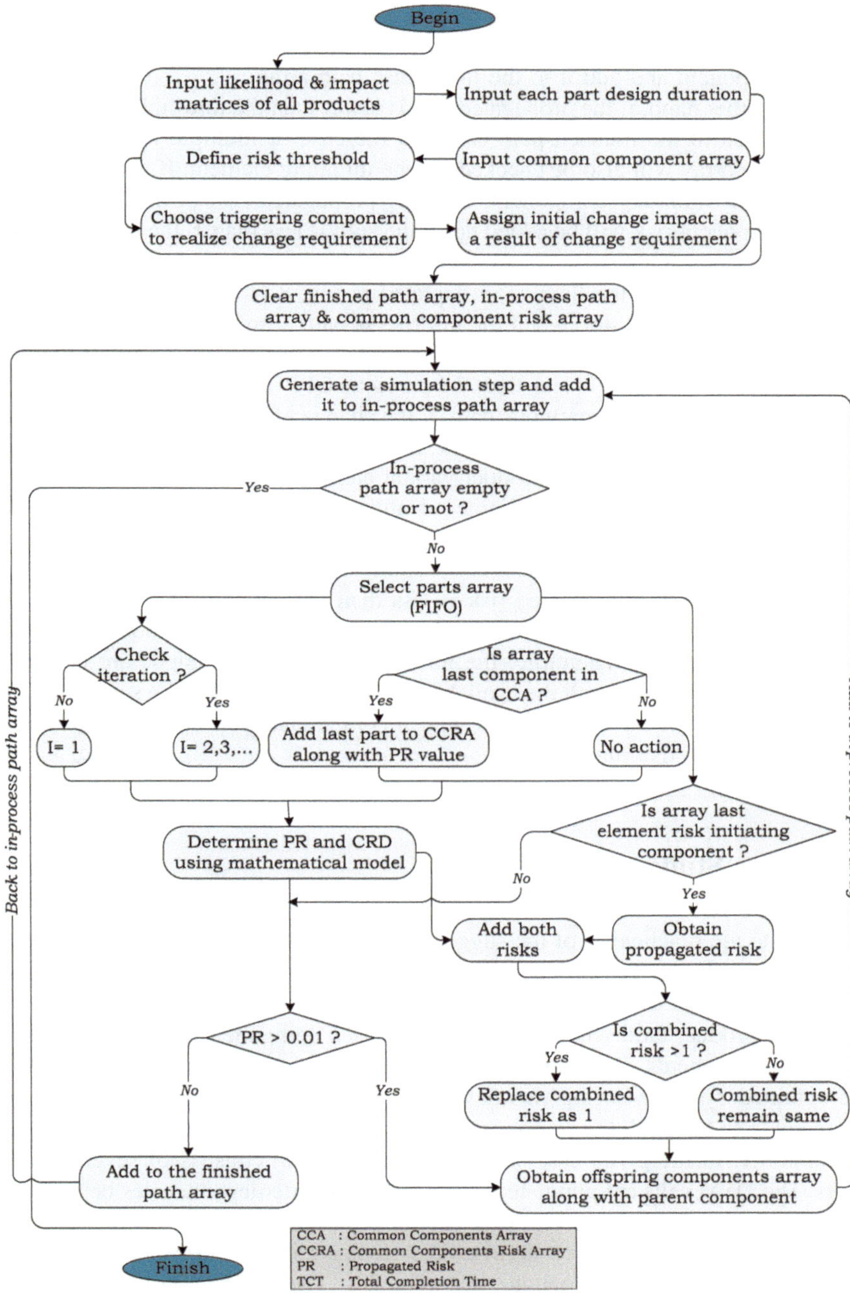

Fig. 8.4 Distinct CPPs flow chart

be calculated according to the mathematical model. If the propagated risk is greater than the threshold value, it will obtain the offspring components along with the parent component and add it to the in-process path array.

On the other hand, if the propagated risk is lower than the threshold value, then it will be added to the finished path array. In the case of multiple instigating components, the array will also be checked for the initiating element. If the last element of an array is initiating component, then the propagated risk of an array and initiating component risk will be added up. If the sum is less than 1, the added risk will remain the same. Otherwise, it will be replaced with 1, and the simulation will continue according to the flow chart depicted in Fig. 8.4.

The simulation will stop when the in-process path array is empty, which means that all the propagation paths have been traced. If the execution time of the subsequent propagation path is less than the previous CPP, then it will be stored in the final array earlier than the previous path. If the change effect of the propagation path is less than 1% of the initial design work, then its effects will not propagate further. As a result, all the CPPs in a product variant can be explored.

The suggested algorithm has some advantages over the existing shortest path algorithms such as Viterbi algorithm [26]. Viterbi algorithm has no stop criterion, which means that it can only discover the shortest path for the finite task. However, in our case when the propagated risk is less than the defined threshold value, the process will stop. Over and above that the Viterbi algorithm does not consider the iteration phenomenon. Therefore, the number of propagation steps are deterministic. In Viterbi algorithm, all the available tasks move forward at the same time. So, the impact difference that constantly happens in the change propagation process is not considered in that algorithm.

8.5 Case Study

To validate the practicality of the suggested framework, a case study of an electric kettle is considered. The product family comprising of four electric kettles (basic kettle, home kettle, office kettle, business kettle) are taken from Simpson et al. [27]. The kettle product family has a total number of ten components. The dependencies between components of product variants and the design duration of each element are obtained by interviewing two design engineers from the kettle manufacturing company. The list of all the components along with their design duration is depicted in Table 8.1. In the given case study, product platform consists of seven components. The DSM cells are populated according to the interdependencies between the components as illustrated in Fig. 8.5. The numeral appears in any off-diagonal cell depicts the direct dependency between the elements in the corresponding *i*th row and *j*th column. Each cell contains two numerals. The number above the line in a cell signifies the change probability, while the number below the line in a cell

Table 8.1 Design component completion time

S. no.	Design component	Design time (h)	Product platform	Abbrev.
1.	Container (whistle)	20	x	CWh
2.	Container (window)	24	√	CW
3.	Container (metal)	22	x	CM
4.	Base (attached)	12	√	BA
5.	Base (detached)	16	√	BD
6.	Control board	18	√	CB
7.	Input panel	10	√	IP
8.	Coiled (bottom)	6	√	CBm
9.	Coiled (side)	10	x	CS
10.	Thermocouple	4	√	Th

denotes the change impact. For example, in the case of product variant basic kettle, the container (whistle) component has 40% chance of propagating changes to the base (attached) component with a change impact of 60% as shown in Fig. 8.5a. The upper and lower bounds for probability and change impact values are from 0 to 1.

8.5.1 Change Propagation Analysis in a Business Kettle

Change propagation in any product variant can take place as a result of change request. Change requirement can be initiated as a customer demand, technological advancement, and internal fault, manufacturing constraints or any legal bounding. The prediction of change and its management in a product family is crucial because it can affect the entire family. For instance, an EC in the business kettle is raised, and the designer investigates that the container (metal) component has to be modified to satisfy the change request. The designer based on experience suggests that the change can cause an initial risk of 0.9 to the container (metal) component. The risk in this chapter is related to the redesign duration of the component. When the change is incorporated in the originating element, it will propagate further to the linked components, thus affecting the whole structure of the product variant. As previously discussed that numerous CPPs as candidates may initiate from an instigating component container (metal), the path with minimum redesign duration is a favorable choice to execute the changes. The total redesign duration of each CPP is plotted as a bar graph, which is illustrated in Fig. 8.6 to analyze the change propagation.

It can be seen in Fig. 8.6 that 51 CPPs are available as a result of the change request. All the paths have different redesign duration depends on the components involved in the change propagation. It is evident from Fig. 8.6 that the path number 18 (green color) has the minimum redesign duration, while 34th CPP (red color) has the maximum redesign duration. The average redesign duration to execute the

Design Components		1	4	8
Container (whistle)	1		0.4 / 0.6	0.3 / 0.5
Base (attached)	4	0.6 / 0.4		0.8 / 0.6
Coiled (bottom)	8	0.8 / 0.2	0.3 / 0.4	

(a) Basic Kettle

Design Components		2	5	6	8	10
Container (window)	2		0.3 / 0.5	0.4 / 0.6	0.3 / 0.5	0.1 / 0.2
Base (detached)	5	0.6 / 0.3				
Control board	6	0.8 / 0.3				
Coiled (bottom)	8	0.8 / 0.2				0.5 / 0.4
Thermocouple	10	0.2 / 0.1			0.4 / 0.3	

(b) Home Kettle

Design Components		2	5	6	7	8	10
Container (window)	2		0.3 / 0.5	0.4 / 0.6	0.4 / 0.7	0.3 / 0.5	0.1 / 0.2
Base (detached)	5	0.6 / 0.3					
Control board	6	0.8 / 0.3			0.6 / 0.7		
Input panel	7	0.8 / 0.4		0.5 / 0.5			
Coiled (bottom)	8	0.8 / 0.2					0.5 / 0.4
Thermocouple	10	0.2 / 0.1				0.4 / 0.3	

(c) Office Kettle

Design Components		3	4	6	7	9	10
Container (metal)	3		0.4 / 0.6	0.4 / 0.6	0.4 / 0.7	0.5 / 0.7	0.2 / 0.2
Base (attached)	4	0.6 / 0.4					
Control board	6	0.8 / 0.3			0.6 / 0.7		
Input panel	7	0.8 / 0.4		0.5 / 0.5			
Coiled (side)	9	0.9 / 0.5					0.5 / 0.4
Thermocouple	10	0.2 / 0.2				0.4 / 0.3	

(d) Business Kettle

Fig. 8.5 Probability and change impact matrices of electric kettle product family

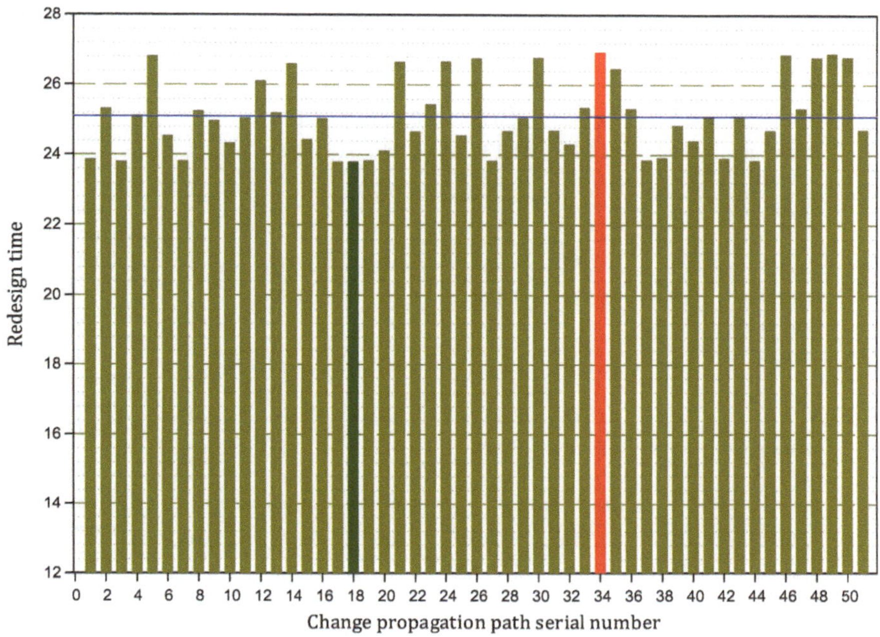

Fig. 8.6 CPPs analysis based on redesign duration in business kettle

change requirement is 25.11 h. 30 out of 51 CPPs have the redesign duration below average. Thus, the CPPs above the mean value of redesign duration should be avoided as far as possible. The detail of fours CPPs is depicted in Table 8.2.

From Table 8.2, it is evident that the redesign duration for instigating component is higher, and it reduces as the change propagates to the other connected component because the propagated risk decreases. The redesign duration for a component is calculated by multiplying the propagated risk to the original design time. For instance, the component 7 (input panel) in a CPP 1 is considered. The propagated risk to component 7 is 0.252 as shown in Table 8.2 and the original design time of component 7 is 10 h as depicted in Table 8.1, then the redesign time of component 7 is 2.52 h ($0.252 \times 10 = 2.52$). The 18th CPP (3-7-6-7-6) is selected to incorporate the design changes in business kettle to satisfy the demand. The CPP is illustrated in Fig. 8.7.

The path comprises of three distinct components, i.e., Container (metal), Input panel, and control board. The number of change steps involves in the change propagation is four, which means that the path must contain five elements. As discussed earlier that three distinct components are participating in the change propagation, which means that some components are iterated in the path. It is evident from Fig. 8.7 that control board and input panel are iterated twice in the CPP. The component 6 (control board) and component 7 (input panel) are product platform; it means that these elements also exist in other product variants.

Table 8.2 Total redesign time of different CPPs initiating from a container (metal) component in a product variant business kettle

1	Path 1	3 → 7 → 6 → 3 → 6						Total redesign time (h) 23.86
	Risk	0.9	0.252	0.063	0.015	0.004		
	Redesign time (h)	19.8	2.52	1.134	0.33	0.072		
2	Path 18	3 → 7 → 6 → 7 → 6						Total redesign time (h) 23.78
	Risk	0.9	0.252	0.063	0.0265	0.0033		
	Redesign time (h)	19.8	2.52	1.134	0.265	0.0594		
3	Path 34	3 → 9 → 3 → 6 → 7 → 3						Total redesign time (h) 26.93
	Risk	0.9	0.315	0.142	0.034	0.0143	0.0046	
	Redesign time (h)	19.8	3.15	3.124	0.612	0.143	0.1012	
4	Path 51	3 → 7 → 3 → 9 → 3 → 6						Total redesign time (h) 24.71
	Risk	0.9	0.252	0.0806	0.028	0.013	0.003	
	Redesign time (h)	19.8	2.52	1.7732	0.28	0.286	0.054	

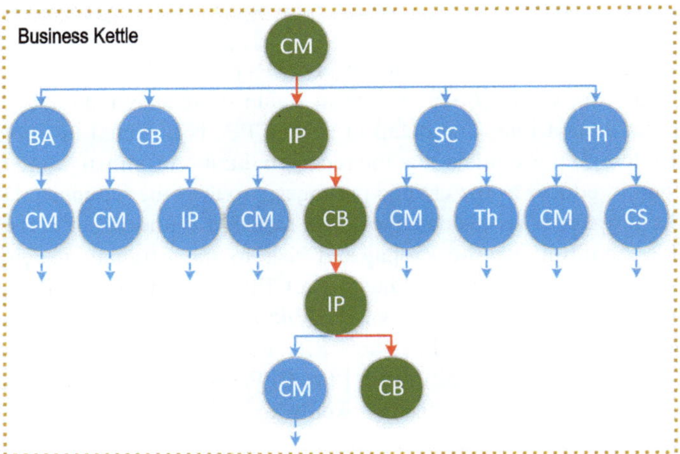

Fig. 8.7 Propagated path in product variant business kettle

Therefore, the changes will further propagate and influence the other products in the family. Component 6 is present in the home kettle product variant while both components 6 and 7 exist in the office kettle product variant. Both these components will serve as a risk originating components in the case of home and office kettle product variants.

8.5.2 *Change Propagation Analysis in an Office Kettle*

The propagation path selected in business kettle product variant contains some common components, i.e., control board and input panel. The common component control board is present in home kettle, and it has a combined propagated risk of 0.0663 (0.063 + 0.0033) as shown in Table 8.2. The product variant office kettle has both the common components and the combined risk propagated to office kettle is 0.3448 [(0.252 + 0.0265) + (0.063 + 0.0033)] as depicted in Table 8.2. The risk propagated to office kettle is more than the risk propagated to the home kettle. Therefore, in the next phase, an office kettle is selected to propagate the changes further. In this case, the risk propagated to both components will be initiated risk. The initiated risk for components 7 and 6 is 0.2785 and 0.0663 respectively.

When the changes are implemented in the office kettle product variant, 24 CPPs are achieved as a result of risk propagation between the elements as displayed in Fig. 8.8. The number of CPPs in the office kettle is less than the number of CPPs in the business kettle. The reason is that the initiated risk in case of the business kettle is much higher than the risk in case of the office kettle. Thus, we can conclude that the number of CPPs depends on the amount of initial risk. It is evident from Fig. 8.8 that 22nd propagation path is the most favorable path and the 4th propagation path is the riskiest path to implement the changes. In this case, only ten propagation paths have redesign duration below the average redesign time. It means

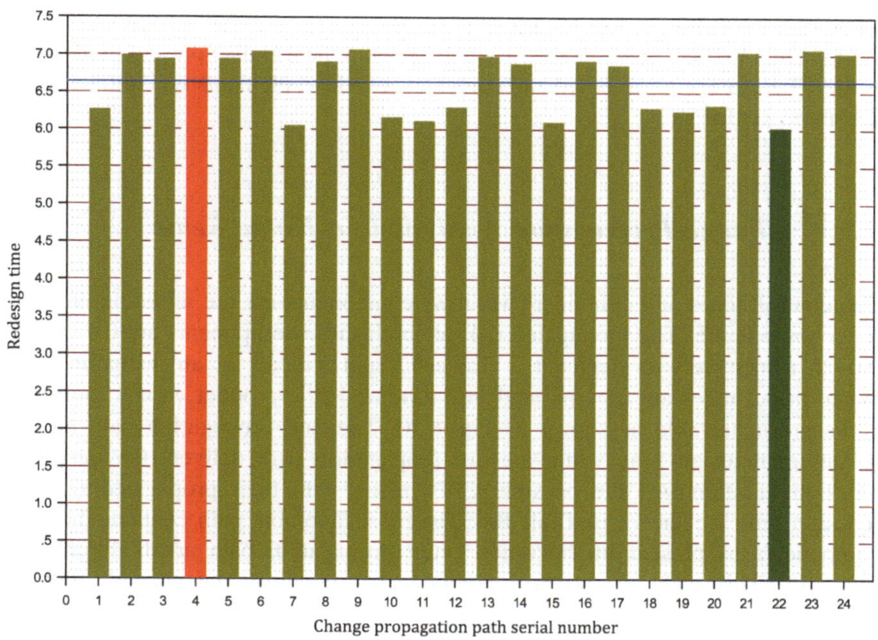

Fig. 8.8 CPPs analysis based on redesign duration in a product variant office kettle

that more than 50% of the CPPs have higher redesign time than the average redesign duration. The designer must be careful to execute the changes in the situation mentioned above. In the case of riskiest CPP, the changes will propagate to the entire system, which can jeopardize the system integrity and can cause an avalanche of the changes.

Four CPPs along with the propagated risk and redesign time of an individual element are illustrated in Table 8.3. The riskiest path has the redesign time of 7.06 h while the most favorable path has the redesign duration of 6.01 h. In the advanced algorithm, the elements are not constrained that each path can traverse. Therefore, the same element can exist in the multiple CPPs because of their small amount of cost for determining design concerns. Consequently, distinct CPPs have small differences in the total redesign duration for the same initial change risk.

The 22nd propagation path (7-6-2-10) in the office kettle product variant is preferred to incorporate the changes because it has minimum redesign time as highlighted in Fig. 8.9. The path contains four distinct components, i.e., an input panel, control board, container (window) and thermocouple. In this path the iteration of any component is not observed because all the elements appear only once in the path. The input panel and control board components have already been considered for redesigning in the business kettle CPP. Therefore, in the office kettle, both these components only serve as a triggering elements. The container (window) and thermocouple components are the new elements in the propagation path. Both these components are part of the product platform. It is evident from Table 8.3 that the propagated risk of thermocouple is 0.0006, which is below the defined threshold value of 0.01. It means that the propagated risk is so small that it cannot affect the components further. The container (window) component is also present in the home kettle. Therefore, control board and container (window) elements will initiate risk in the home kettle.

8.5.3 Change Propagation Analysis in a Home Kettle

The control board and container (window) components are the initiating elements for a home kettle. The propagated risk for both these components are 0.1359 and 0.0326 respectively as depicted is Table 8.3 and will act as an instigating risk. When the changes are incorporated in the home kettle product variant, only four CPPs are achievable as shown in Fig. 8.10. The 2nd propagation path has the minimum redesign time while the 3rd propagation path delivers the maximum redesign duration. All the possible CPPs along with their propagated risk and redesign time are shown in Table 8.4. The most favorable path contains the same components as present in the office kettle propagation path are highlighted in Fig. 8.11. This path has no new components; therefore, the changes will not propagate further.

Table 8.3 Total redesign time of different CPPs initiating from input panel and control board components in a product variant office kettle

1	Path 1	7 → 6 → 7 → 2 → 8					Total redesign time (h) 6.26
	Risk	0.278	0.1359	0.0571	0.0183	0.0027	
	Redesign time (h)	2.78	2.446	0.571	0.439	0.0162	
2	Path 4	7 → 2 → 6 → 2 → 7					Total redesign time (h) 7.06
	Risk	0.278	0.089	0.0877	0.021	0.0059	
	Redesign time (h)	2.78	2.136	1.5786	0.504	0.059	
3	Path 22	7 → 6 → 2 → 10					Total redesign time (h) 6.01
	Risk	0.278	0.1359	0.0326	0.0006		
	Redesign time (h)	2.78	2.446	0.7824	0.0024		
4	Path 24	7 → 2 → 6 → 2 → 10					Total redesign time (h) 7.00
	Risk	0.278	0.089	0.0877	0.021	0.0004	
	Redesign time (h)	2.78	2.136	1.5786	0.504	0.0016	

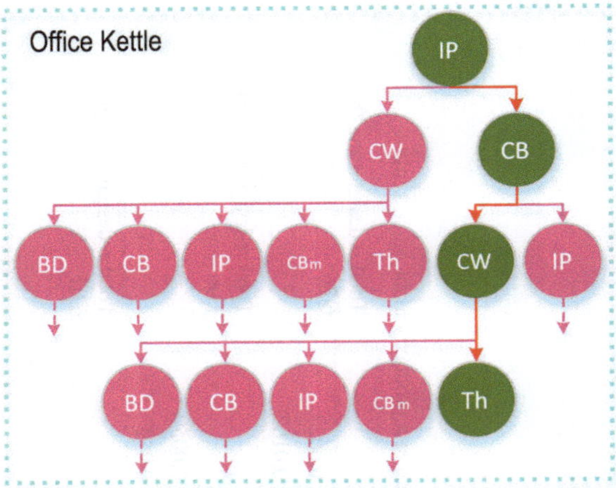

Fig. 8.9 Propagated path in a product variant office kettle

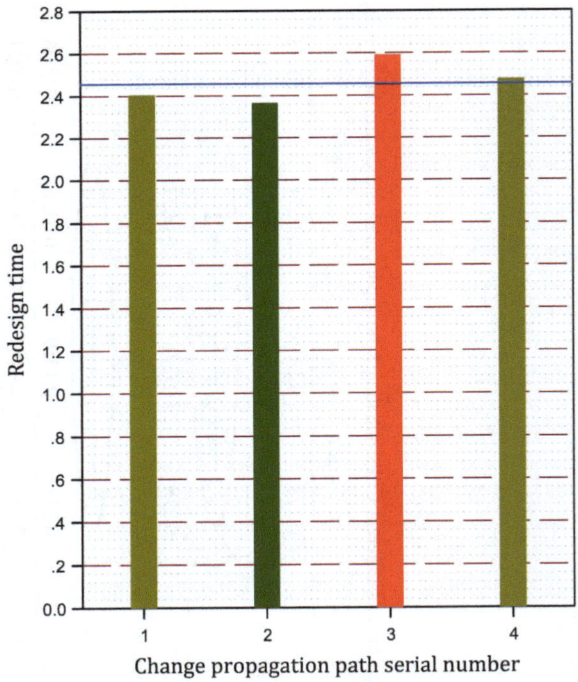

Fig. 8.10 CPPs analysis based on redesign duration in home kettle

Table 8.4 Total redesign time of different CPPs initiating from control board component in a product variant home kettle

1	Path 1	6 → 2 → 8				Total redesign time (h) 2.40
	Risk	0.0663	0.0485	0.0073		
	Redesign time (h)	1.1934	1.164	0.0438		
2	Path 2	6 → 2 → 10				Total redesign time (h) 2.36
	Risk	0.0663	0.0485	0.001		
	Redesign time (h)	1.1934	1.164	0.004		
3	Path 3	6 → 2 → 6 → 2				Total redesign time (h) 2.58
	Risk	0.0663	0.0485	0.0116	0.0008	
	Redesign time (h)	1.1934	1.164	0.2088	0.0192	
4	Path 4	6 → 2 → 5				Total redesign time (h) 2.47
	Risk	0.0663	0.0485	0.0073		
	Redesign time (h)	1.1934	1.164	0.1168		

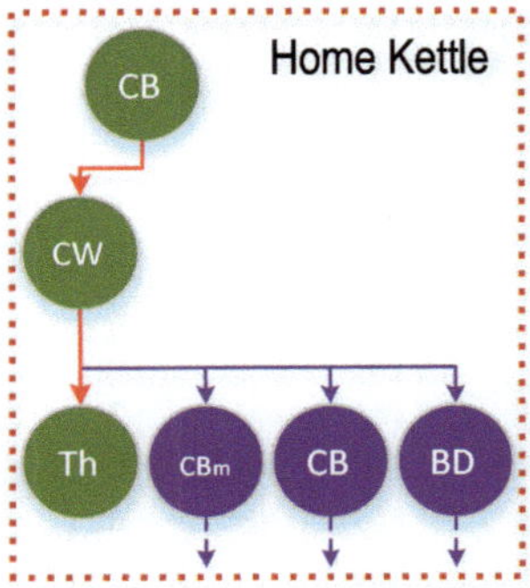

Fig. 8.11 Propagated path in a product variant home kettle

8.5.4 Change Propagation Analysis in a Product Family Design of Electric Kettle

The results obtained due to the application of proposed technique are now consolidated in this sub-section. Product family consists of four product variants having 10 components. Seven elements are the part of a product platform, and three are sole components. The initiated change is propagated to three product variants except basic kettle because of change requirement. The most favorable CPPs in three product variants are picked to execute the changes as shown in Table 8.5. It can be seen from Table 8.5 that some components appear repeatedly in distinct CPPs. It is observed that five distinct components from different product variants are involved in the change propagation. The components having maximum risk in the CPPs are selected to calculate the CRD. This risk is multiplied by the original design time of each element and then add all the outcomes to get the redesign time of product family as shown in Table 8.6. The overall CPP in three product variants is provided in Fig. 8.12.

The list of the components from different product variants, which are involved in the change propagation, is depicted in Table 8.7. From the literature, it is revealed that the change can usually propagate up to five follow-up changes arising from instigating component [28]. Thus, the maximum number of elements available in a CPP can be up to 6. It is evident from Tables 8.2, 8.3 and 8.4 that the outcomes of our approach are in line with the work performed by Pasqual and de Weck [28].

Table 8.5 Favorable CPPs in PF of electric kettle

Business kettle	Path 18	3 → 7 → 6 → 7 → 6				
	Risk	0.9	0.252	0.063	0.0265	0.0033
Office kettle	Path 22	7 → 6 → 2 → 10				
	Risk	0.2785	0.1359	0.0326	0.0006	
Home kettle	Path 2	6 → 2 → 10				
	Risk	0.0663	0.0485	0.001		

Table 8.6 CRD to incorporate changes in PF of electric kettle

Distinct change components			
Component	Max risk in CPP	Initial design duration (h)	Redesign duration (h)
3	0.9	22	19.8
7	0.2785	10	2.785
6	0.1359	18	2.4462
2	0.0485	24	1.164
10	0.001	4	0.004
CRD			26.2

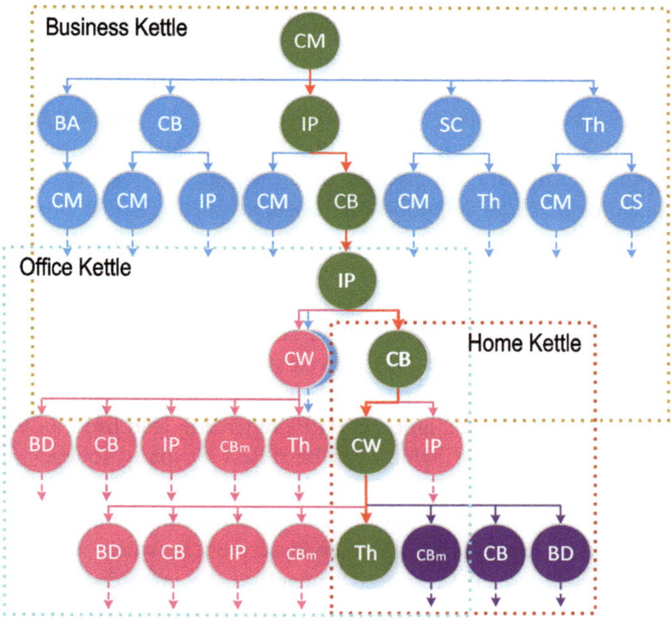

Fig. 8.12 Propagated path in a product family of electric kettle

Table 8.7 List of design components modified as a result of change propagation

S. no.	Design component	Product variant
1.	Container (window)	Home and office kettle
2.	Container (metal)	Business kettle
3.	Control board	Home, office and business kettle
4.	Input panel	Office and business kettle
5.	Thermocouple	Home, office and business kettle

From the designer point of view, the path which contains a fewer number of design components and change step will be easier and require minimum time to implement the changes. The CRD of change propagation across product family is 26.2 h. On the other hand, 34th CPP in the business kettle product variant has the redesign time of 26.93, which is 0.73 h more than the overall redesign time of product family. Therefore, based on the above, it can be concluded that the selection of an appropriate CPP in product family is critical to avoid any delay in the project.

8.6 Conclusion and Future Work

In recent years, platform-based PD has gained much consideration owing to design reuse and low variable manufacturing cost. The main focus of product family design is to quickly deliver products with a great variety to meet the needs of potential customers and capture new market niches. In product family design, considerable progress has been made in the areas of designing, planning, manufacturing, commonality, and modularity of the components except EC prediction and its management. Little attention has been given to change propagation in a product family. EC execution in a product family affects not only the product being under consideration, but also other product variants in the family. Most of the existing change propagation evaluation techniques are limited to a single product, and they do not effectively support product family design.

This chapter introduced an advanced seven steps change propagation mechanism to predict and evaluate change propagation across product variants in a family. The main emphasis is to foresee the efficient CPPs regarding lead time. In the first step, the change request is initiated, which can be raised by many sources as discussed earlier. In the second step, all the product variants are decomposed into elements and then interdependencies between the components are investigated in the third step. Logical relationships '*AND/OR*' are described between the product's elements. Component-based DSM is used to capture the interdependencies between the directly connected elements. In the fourth step, an appropriate product variant is selected to implement the change request using a proposed algorithm and mathematical model. The algorithm has been made by doing some alterations in the BFS technique, which is used to explore the shortest path. The suggested algorithm is based on the quantitative risk analysis approach to evaluate the propagated risk between the elements. Numerous CPPs along with the redesign time and propagated risk are achieved. In step five, the CPP with a minimum redesign time is selected, and the common components (product platform) are highlighted. After that, the changes are propagated to that product variant, whose common components have maximum propagated risk. In step six, the CPPs are explored in other product variants until the propagated risk reaches to its threshold value. In step seven, the CRD is calculated by adding the redesign time of each product variant involved in the change propagation.

A case study, based on the family of four electric kettles is presented in this chapter. Risk analysis is performed at the component level to predict the change propagation in a family. The results exhibit that numerous CPPs are available in each product variant to execute the changes. The number of CPPs in each product variants decreases as the change propagates. It can be noticed that the CRD of product family is 0.73 h less than the redesign time of an individual propagation path in business kettle product variant. Thus, it can be concluded that the selection of suitable CPP in a family of product is essential to avoid any delay in the project. The results revealed that some components appeared two or three times in the propagation path because of an iteration phenomenon. Consequently, the case study exhibits the usefulness of proposed methodology for the prediction of changes in product family and provides an opportunity for a company to deliver its product without delays. The authors, however, acknowledge that more work is also needed to validate the modeling technique further. The suggested approach will be tested in more case examples. Future research will be focused, to extend the change propagation from product family to the related tooling.

References

1. Simpson TW (2004) Product platform design and customization: status and promise. Artif Intell Eng Des Anal Manuf 18(1):3–20. doi:10.1017/s0890060404040028
2. Khajavirad A, Michalek JJ, Simpson TW (2009) An efficient decomposed multiobjective genetic algorithm for solving the joint product platform selection and product family design problem with generalized commonality. Struct Multidiscip Optim 39(2):187–201. doi:10.1007/s00158-008-0321-9
3. Meyer MH, Lehnerd AP (1997) The power of product platforms: building value and cost leadership. Free Press, New York
4. McGrath ME (1995) Product strategy for high-technology companies. Irwin Professional Publishing, New York
5. Marion TJ, Thevenot HJ, Simpson TW (2007) A cost-based methodology for evaluating product platform commonality sourcing decisions with two examples. Int J Prod Res 45 (22):5285–5308. doi:10.1080/00207540600710970
6. Park J, Simpson TW (2008) Toward an activity-based costing system for product families and product platforms in the early stages of development. Int J Prod Res 46(1):99–130. doi:10.1080/00207540600825240
7. Ahmad N, Wynn DC, Clarkson PJ (2013) Change impact on a product and its redesign process: a tool for knowledge capture and reuse. Res Eng Des 24(3):219–244. doi:10.1007/s00163-012-0139-8
8. Hamraz B, Clarkson PJ (2015) Industrial evaluation of FBS linkage—a method to support engineering change management. J Eng Des 26(1–3):24–47. doi:10.1080/09544828.2015.1015783
9. Koh ECY, Caldwell NHM, Clarkson PJ (2012) A method to assess the effects of engineering change propagation. Res Eng Des 23(4):329–351. doi:10.1007/s00163-012-0131-3
10. Van Bossuyt DL, Dong A, Tumer IY, Carvalho L (2013) On measuring engineering risk attitudes 1. J Mech Des 135(12):121001/1–121001/13. doi:10.1115/1.4025118

11. Zhang J, Song X, Chen H, Shi R (2015) Optimisation of critical chain sequencing based on activities' information flow interactions. Int J Prod Res 53(20):6231–6241. doi:10.1080/00207543.2015.1043157
12. Jiao J, Simpson TW, Siddique Z (2007) Product family design and platform-based product development: a state-of-the-art review. J Intell Manuf 18(1):5–29. doi:10.1007/s10845-007-0003-2
13. Berry S, Pakes A (2007) The pure characteristics demand model. Int Econ Rev 48(4):1193–1225. doi:10.1111/j.1468-2354.2007.00459.x
14. Wortmann JC, Muntslag DR, Timmermans PJM (1997) Customer-driven manufacturing, 1st edn. Springer, Chapman & Hall, Amsterdam, London. doi:10.1007/978-94-009-0075-2
15. Tseng MM, Jiao J, Merchant ME (1996) Design for mass customization. CIRP Ann Manuf Technol 45(1):153–156. doi:10.1016/S0007-8506(07)63036-4
16. Lee H, Seol H, Sung N, Hong YS, Park Y (2010) An analytic network process approach to measuring design change impacts in modular products. J Eng Des 21(1):75–91. doi:10.1080/09544820802232517
17. Clarkson PJ, Simons C, Eckert C (2004) Predicting change propagation in complex design. J Mech Des 126(5):788–797. doi:10.1115/1.1765117
18. Eckert C, Clarkson PJ, Zanker W (2004) Change and customisation in complex engineering domains. Res Eng Des 15(1):1–21. doi:10.1007/s00163-003-0031-7
19. Keller R, Eckert CM, Clarkson PJ (2009) Using an engineering change methodology to support conceptual design. J Eng Des 20(6):571–587. doi:10.1080/09544820802086988
20. Yang F, Duan G-J (2012) Developing a parameter linkage-based method for searching change propagation paths. Res Eng Des 23(4):353–372. doi:10.1007/s00163-011-0124-7
21. Fei G, Gao J, Owodunni O, Tang X (2011) A method for engineering design change analysis using system modelling and knowledge management techniques. Int J Comput Integr Manuf 24(6):535–551
22. Kocar V, Akgunduz A (2010) ADVICE: a virtual environment for engineering change management. Comput Ind 61(1):15–28. doi:10.1016/j.compind.2009.05.008
23. Raffaeli R, Malatesta M, Marilungo E, Germani M (2013) An approach for managing engineering changes in product families. In: Proceedings of the ASME, international design engineering technical conferences, Portland, Oregon, pp 1–12
24. Steward DV (1981) The design structure system: a method for managing the design of complex systems. IEEE Trans Eng Manag EM-28(3):71–74. doi:10.1109/TEM.1981.6448589
25. Browning TR (2016) Design structure matrix extensions and innovations: a survey and new opportunities. IEEE Trans Eng Manag 63(1):27–52. doi:10.1109/TEM.2015.2491283
26. Viterbi J (1967) Error bounds for convolutional codes and an asymptotically optimum decoding algorithm. IEEE Trans Inf Theory 13(2):260–269
27. Simpson TW, Jiao JR, Siddique Z, Hölttä-Otto K (2014) Advances in product family and product platform design: methods and applications, 1st edn. Springer, New York. doi:10.1007/978-1-4614-7937-6
28. Pasqual MC, de Weck OL (2012) Multilayer network model for analysis and management of change propagation. Res Eng Des 23(4):305–328. doi:10.1007/s00163-011-0125-6

MIX
Papier aus verantwortungsvollen Quellen
Paper from responsible sources
FSC® C105338

If you have any concerns about our products,
you can contact us on
ProductSafety@springernature.com

In case Publisher is established outside the EU,
the EU authorized representative is:
Springer Nature Customer Service Center GmbH
Europaplatz 3, 69115 Heidelberg, Germany

Printed by Libri Plureos GmbH
in Hamburg, Germany